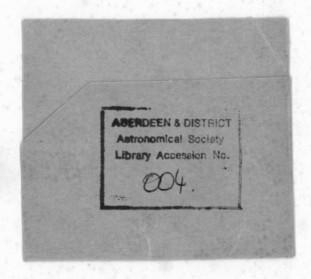

A TEXT BOOK
OF
ELEMENTARY ASTRONOMY

Total eclipse of the Sun: Aegean Sea, 1936

A TEXT BOOK
OF
ELEMENTARY ASTRONOMY

By

ERNEST AGAR BEET

B.Sc., F.R.A.S.

CAMBRIDGE
AT THE UNIVERSITY PRESS
1953

PUBLISHED BY
THE SYNDICS OF THE CAMBRIDGE UNIVERSITY PRESS

London Office: Bentley House, N.W. 1
American Branch: New York

Agents for Canada, India, and Pakistan: Macmillan

First Edition 1945
Reprinted 1946
Reprinted with Corrections 1953

Printed in Great Britain at the University Press, Cambridge
(Brooke Crutchley, University Printer)

CONTENTS

Chapters

ILLUSTRATIONS

PREFACE

A WIDER OUTLOOK in school science has been growing for some years, but although Astronomy is a subject often recommended it is seldom taught. The objections to Astronomy are, presumably, that it does not provide suitable experimental work and that time cannot be spared in an already overcrowded time table. With the rapid increase of General Science it is to be hoped that this section will eventually find its place, and this book more than covers the syllabus suggested by the Science Masters' Association.[1]

The book may also have an appeal outside the schools, as some may wish for a course of instruction more formal than the general reading already well catered for in the extensive literature of the subject. It will form a sequel to my former book, *A Guide to the Sky*, which is an observational introduction for young people. Astronomy may be approached from a mathematical or experimental standpoint. The former has already been ably done by P. F. Burns in his *First Steps in Astronomy*; an experimental and historical approach is made here.

Simple experimental work which forms a part of the main argument appears in its place in the text, as do some suggested demonstrations. At the ends of the chapters there will be found other exercises and out-door work of which the importance cannot be overstressed. The questions at the end of the book are graded, A being preliminary questions intended to direct the thoughts before reading the chapter, B questions on the text, and C of a problem nature.

Figs. 21, 67, 68, 71, 72, 76, 78, 79, 81, 87 and 90 are to be found in *The Stars in their Courses* by Sir James Jeans; Fig. 89 is in the same author's *The Universe Around Us*; Figs. 61 and 64 are reproduced from *Light* by A. E. E. McKenzie. For permission to use these the author is grateful to the Cambridge University Press, and to the owners of the copyrights to whom ascription is made on the figures themselves.

Acknowledgements are also tendered to Dr J. L. Haughton, F.R.A.S., and the British Astronomical Association for the frontispiece; Sir Howard

---

[1] *The Teaching of General Science*, Part II, 1938.

Grubb Parsons and Co. for Figs. 57, 58 and 63; the Director of the Yerkes
Observatory for Fig. 88; The *New York Times* for Fig. 65; Mr P. M. Ryves,
F.R.A.S., and the family of the late T. E. R. Phillips for Fig. 70; Mr F. J.
Sellers, F.R.A.S., for Question 72; Commander W. S. MacIlwaine, R.N.,
for reading and criticising a part of the manuscript; my father, Rev.
Dr W. E. Beet, F.R.Hist.S., for reading the proofs; and to many authors
and friends whose ideas on teaching Astronomy have influenced mine
and have thus become incorporated in this book.

E. A. B.

NAUTICAL COLLEGE,
PANGBOURNE.

December 1944

CHAPTER I

CONCERNING LIGHT

SIGHT is a very wonderful sense indeed. Touch and taste demand actual contact with something if they are to give us any information about it. Our sense of smell can sometimes be used over a short distance, and hearing may even be used over some miles. Sight shows us things and events at distances almost too great to imagine. In a pitch-dark room you see nothing, for sight depends upon light. When a match is struck it somehow causes disturbances called light waves and these, like water waves from a fallen stone, spread out in all directions, until they meet your eye. The eye is sensitive to light, the optic nerve communicates the impression to the brain, and you become aware that in a certain place a match is burning.

This book is mainly concerned with other worlds than ours, bodies very many thousands of miles away, and our knowledge of them comes to us as light messages affecting our sense of sight. Thus a study of astronomy is closely allied with a study of light, and it is therefore desirable to become familiar with the chief properties of the latter before proceeding to the former.[1]

It is necessary to be reminded, right at the beginning, of one important property. That we cannot see around corners is a common experience, and the reason is that the light by which we see travels in straight lines. Notice the beam of light entering a dim and dusty room, or a motor-car headlight on a slightly misty night, or the sunlight shining through a rift in the clouds, and there will be no doubt about the reality of these straight lines. Thus when we want to show on a diagram the direction in which light is travelling we just rule a straight line and call it a 'ray', and to show the direction in which we can see we rule such a line

[1] Readers who have already done a course of light up to the standard of the General Certificate may skip the rest of this chapter if they wish, but those who have not are advised not only to persevere with it but to read also a text book on Light. *Light*, by A. E. E. McKenzie (Cambridge University Press), can be recommended as being in sympathy with the present work.

and call it the 'line of sight'. It is therefore important to remember that *light travels in straight lines.*

Reflection. When light falls on any ordinary surface, such as the page of this book, it is reflected in all directions, perhaps not quite equally, but sufficiently so for the surface to be seen from any angle. When it falls on a highly polished one, such as a mirror, it is reflected in one direction, and the *laws of reflection* are best understood by trying two simple experiments.

Expt. 1. Project a narrow streak of light across a sheet of paper by putting a lamp behind a vertical slit in a piece of cardboard, or shining it between two rectangular objects.[1] Let it strike a plane (i.e. flat) mirror standing up at right angles to the paper. This ray *AB* (Fig. 1) is the *incident ray*, and the ray *BC* is the *reflected ray*. Rule lines along them, and along the back of the mirror if a glass one or the front if a metal one. Draw *BN* at right angles to the mirror; it is called the *normal*. The angles *ABN* and *NBC* are called the *angle of incidence* and *angle of*

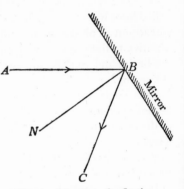

Fig. 1. The laws of reflection

reflection; measure them with a protractor, and if you have done your experiment carefully they will be found to be equal.

Expt. 2. Now tilt the mirror slightly so that it is no longer at right angles to the paper; what happens to the reflected ray? The paper on which you are working is a plane, and the incident and reflected rays lie in that plane. So does the normal, because it is perpendicular to the mirror and the mirror is perpendicular to the paper. When the mirror is tilted the normal is moved out of the plane, and you find that the reflected ray moves too and disappears.

[1] Simple methods of projecting rays will be found in *The Science Masters' Book,* Series 1, Vol. 1.

Thus the laws of reflection are (i) the incident and reflected rays and the normal to the surface are all in the same plane, and (ii) the angle of reflection is equal to the angle of incidence.

The reflection of an object in a mirror is called its *image*, and, as it is behind the glass and the light has not really passed through it, it is said to be *virtual*. It can be shown quite easily that the image appears to be, not on the glass, but as far behind the plane mirror as the object is in front.

Have you ever stood in front of a mirror and shaken hands with yourself? You use the right hand and your image uses the left, and this is another rule about a plane mirror. This changing over of right and left is called *lateral inversion*.

Expt. 3. Set up Expt. 1 and rule along the incident and reflected rays and along the mirror. In Fig. 2 *MB* is the position of the mirror and *ABC* the path of the light. Turn the mirror through a small angle into a new position *NB* and mark the new position *BD* of the reflected ray. Measure the angles *NBM* and *CBD*. You will find that the ray has moved twice as much as the mirror, a fact that will occur in a later chapter in connection with the sextant.

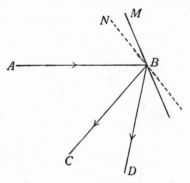

FIG. 2. Effect of rotating a mirror

A *concave mirror* is a part of a sphere, reflecting on the inside of the curve, and those used in telescopes are not sharply curved, like half a tennis ball, but are nearly flat. Again it is easier to understand its properties if you can handle one, so, if such a mirror is available, here are two more simple experiments.

Expt. 3. Place the mirror half way through a slot in a piece of white card, the mirror being perpendicular to the card, and project on to it several rays by passing the light through several slits or a coarse comb. Note that the reflected rays meet at one point; if the incident rays are parallel (Fig. 3) the point is called the *principal focus*, and its distance from the mirror its *focal length*.

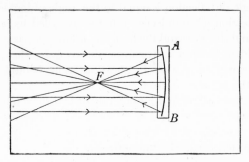

FIG. 3. A concave mirror

Expt. 4. Hold the mirror facing a window, and hold a piece of card so that the reflected light falls upon it. Vary the distance and you will find a position in which an inverted reproduction of the window frame appears on the card. This is called a *real image*, as the light really does fall on the screen. Now hold the mirror close to your face: as the object (the face) is very near the mirror the image is now upright and virtual.

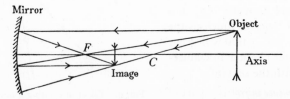

FIG. 4. Formation of a real image by a concave mirror

The formation of a real image can be illustrated by a drawing, for light rays can be represented by straight lines. In Fig. 4 C is the centre of the circle of which the mirror forms a part, and the principal focus F lies half way between C and the mirror. From the top of the object a ray is drawn parallel to the axis; the reflected ray will pass through the focus. A ray drawn through the focus will reflect parallel to the axis. A ray through the centre of curvature will be at right angles to the mirror and will therefore reflect back along its own path. The three rays meet at a point; this is the head of the image.

Refraction. Have you ever noticed any of these things: the appearance of your fingers through a filled tumbler in your hand; a flight of steps leading down into a swimming pool; the bottom of a clear stream seen from a boat; someone moving in the garden seen through very ordinary window glass? If so, you are already familiar with the subject of this section, for the various effects referred to are due to *refraction*, which is the bending of light as it passes from one transparent medium into another.

Expt. 5. Put a coin at the bottom of a basin, move away until the coin is *just* hidden by the edge of the basin, and then watch carefully while a friend pours water in without disturbing the coin.

The reason for the reappearance of the coin is that the light reflected from it is bent when it leaves the water and enters your eye as if it had come from B instead of A (Fig. 5). Thus you see the coin apparently at B. It is important to note that light bends away from the normal on leaving the more dense water and entering the less dense air; it

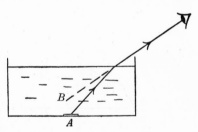

Fig. 5. Refraction in water

would bend towards the normal on entering a more dense medium.

The laws of refraction are not quite so simple as those of reflection, and the reader is referred to text books on light. There are a few facts about refraction that should be noted, however. When light passes through a parallel slab of glass it emerges parallel to its original direction but *laterally displaced* (Fig. 6). If

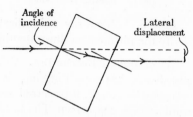

Fig. 6. Refraction through a glass slab

the glass is thin and the angle of incidence is small, this displacement is also small. In the case of a triangular prism the light will be *deviated* towards the wider part of the prism (Fig. 7), the angle D being called the angle of deviation. If the angle of incidence inside the glass is large,[1] i.e. very oblique incidence, refraction does not occur and the light is *internally reflected*, shown in Fig. 8. This internal reflection is important, as in the construction of optical instruments

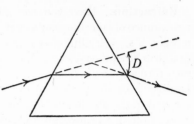

FIG. 7.   Refraction through a prism

right-angled isosceles prisms are frequently used instead of mirrors; one way of using such a prism is shown in Fig. 9.

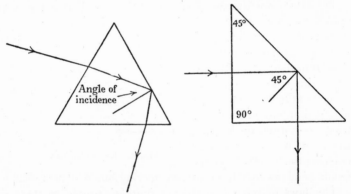

FIG. 8.   Internal reflection   FIG. 9.   A reflecting prism

Expt. 6. Using a ray and a sheet of paper as in Expt. 1, try to verify the phenomena illustrated in Figs. 6–8.

Atmospheric Refraction. Although our study of astronomical phenomena has not yet begun, this is a convenient point at which to insert this topic; if preferred, the paragraph could be omitted until reading chapter VII. The Earth's atmosphere does not go on for ever; it is a layer something over 100 miles thick, and beyond

---

[1] 42° and above for ordinary glass.

it is empty space. When light from the Sun and stars enters the atmosphere it is refracted, because it is entering a denser medium, and this affects astronomical observations. When light from a star S (Fig. 10) enters the atmosphere it reaches an observer at O as if the star were at A. Then the altitude as measured by a sextant would be HOA instead of the true altitude HOB, and therefore sextant readings have to be corrected for refraction. Similarly when the Sun is in the direction of C, just below the horizon, it appears to be at D, just above the horizon; thus refraction lengthens the day. Again, during an eclipse of the Moon some

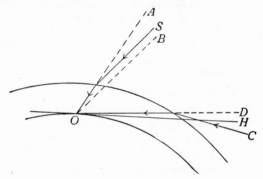

FIG. 10. Atmospheric refraction

light is refracted by the atmosphere into the Earth's shadow and illuminates the Moon enough to make it visible. Finally, refraction is partly responsible for morning and evening twilight. Needless to say, the atmosphere does not begin suddenly; the density of it becomes lower as height increases. Hence the light rays do not make one sharp turn as shown in Fig. 10, but follow a curved path, being only slightly deviated when they first enter the very rare upper layers.

Lenses. Most people are familiar with the common magnifying glass; it is a convex or converging lens, thicker in the middle than at the edge.

Expt. 7. Repeat Expt. 3, using a convex lens in the slot. Note that the refracted rays meet at a point beyond the lens. The terms principal focus and focal length have the same meaning as in the case of the concave mirror (Fig. 11).

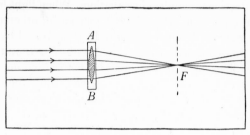

FIG. 11.   A convex lens

Expt. 8. Repeat Expt. 4 with the lens. This time a real image will be obtained by holding your screen on the side of the lens away from the window.

Light from a very distant object can be regarded as being parallel; thus a real image of the Sun will be formed at the principal focus, and this gives a simple method of measuring focal length. Strictly speaking, the 'hot spot' given by a burning glass is not a point, but a disc of measurable size formed in the *focal plane*; this plane is shown dotted in Fig. 11.

The formation of a real image is illustrated in Fig. 12, which is drawn as follows: a ray parallel to the axis will refract through

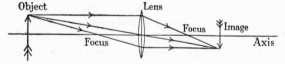

FIG. 12.   Formation of a real image by a convex lens

the focus on the far side; one through the near focus will refract parallel to the axis; one through the centre of the lens will go straight on, for here the lens acts like a thin parallel plate.

When a convex lens is being used as a magnifying glass the object, like that in the mirror case previously mentioned, is so close to the lens that it is within its focal length, and a virtual image is given. This is illustrated in Fig. 13, which is drawn according to the same rules as Fig. 12 except that the rays had to be produced backwards to locate the virtual image.

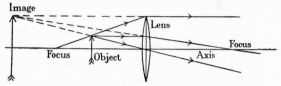

FIG. 13. Formation of a virtual image by a convex lens

A *concave lens* is thinnest in the middle and it *diverges* the light instead of converging it. The only kind of image that it can give by itself is a small, upright and virtual one.

Dispersion. If a triangular glass prism be placed in the path of direct sunlight shining into the room, a patch of coloured light can be obtained on the wall or ceiling. The light is said to have been *dispersed* into a *spectrum*. The colours can be more conveniently examined if a strong source of light, such as a motor-car bulb, with a straight vertical filament, be placed behind a narrow vertical slit so that a narrow beam of light falls on a 60° prism (Fig. 14).

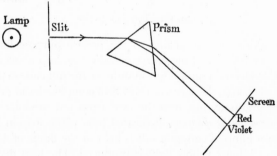

FIG. 14. The dispersion of light

An important point to realise is that the prism does not actually make the colours. The colours are already in the light, and they differ from one another in the same way as the wireless transmission from different stations. Light waves were mentioned in the opening paragraph; each colour has its own particular wave-length, and when the waves fall on a prism the deviation produced depends on the wave-length. White light is a mixture of many wave-lengths, and the prism, by turning each wave-length through a different angle, sorts them out and we can see the colours separately. An experiment of this kind was performed by Sir Isaac Newton in 1666, using sunlight from a hole in a shutter, and is described in his *Opticks*. The colours of the spectrum usually quoted are red (the least deviated), orange, yellow, green, blue, indigo (deep blue) and violet. *Spectroscopes* and *spectrometers* are instruments for producing and examining spectra; their design

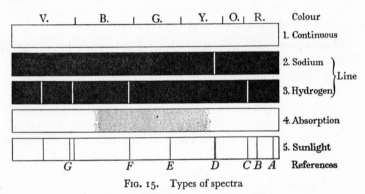

FIG. 15. Types of spectra

varies considerably according to the precise purpose for which they are to be used. If the source of light is an incandescent solid, such as an electric lamp, a gas mantle, or the hot carbon particles in a candle flame, the spectrum will be a complete band of colour, merging gradually one into the next from red to violet. This is called a *continuous spectrum*, illustrated by a white strip in Fig. 15.

When a lump of common salt is held in the flame of a Bunsen burner an intense yellow light is produced. The heat divides up the salt into the elements of which it is composed, and at the temperature of the flame the sodium vapour produced emits a yellow light. If this light be examined with a spectroscope, in place of a continuous spectrum there is just one yellow line (really two close together) as shown in Fig. 15, 2. When an electric current is passed through a gas at a low pressure, light is produced, the Neon signs outside shops being a common example of this sort of lamp. This kind of light also gives lines in a spectroscope, those due to hydrogen being illustrated. There are two very important points about these *bright line spectra*: (*a*) they are caused by incandescent gases or vapours, not by solids; and (*b*) they are characteristic of the element causing them, every element having its own particular group of lines. Chemists can therefore use the spectroscope as a means of detecting elements.

Suppose that a spectrum is being projected on to a screen, and then a piece of red glass be put in front of the slit. Will the spectrum turn red? If you can try the experiment you find that it will not, but that except for the red part the spectrum will disappear. All colours except red have been absorbed, and the result on the screen will be the *absorption spectrum* of that particular piece of glass. All transparent substances have their absorption spectra, that of a fairly dilute solution of potassium permanganate being illustrated in Fig. 15, 4.

The spectroscope is a very powerful tool in the hands of the astronomer, but an account of his use of it will be deferred until we are discussing the Sun in chapter XIV; Fig. 15, 5 will be explained there.

When a spectrum is photographed it is found that the photograph extends beyond the violet end of the visible spectrum. This shows the existence of radiation called the *ultra-violet*, which has a chemical effect on the plate or film, but is not visible to the eye. Similarly, with a special kind of plate, an extension of the spectrum at the red end can be photographed; this is called the *infra-red*. Infra-red photography is a comparatively new thing and is of growing importance as a means of seeing through obscuring mists and over great distances.

When light passes through a lens it is dispersed a little, and thus bright images due to a simple lens are spoilt by having coloured fringes around them. Fortunately this difficulty can be largely overcome by using a combination of convex and concave lenses, of different kinds of glass, in contact. This arrangement is called an *achromatic lens*, and is the kind used as the object glass of a good quality telescope.

CHAPTER II

THE EARTH

WHAT is the shape of the earth? Your answer will probably be 'round', because you have been told so or have read about it in geography. Do you believe it? Does it look round? Except for the various undulations that we call hills and mountains it looks flat rather than round, and for many centuries in the past people regarded it as such. It is so large, and we are so small, that we cannot see its shape directly, but if we consider carefully several things that we can see we realise that it must be round. Some of this evidence that the earth is round you may have seen in your geography books, but there is no harm in looking at it again.

(i) It is possible to travel right around the world. This should need no further explanation.

(ii) Ships in the distance are described as being 'hull down on the horizon'. That is, only the masts and funnels are visible, and as they come nearer more of them comes in sight. In Fig. 16 the

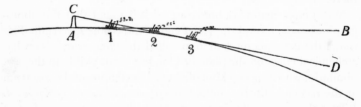

FIG. 16. The curvature of the Earth

curved line is the surface; *AB* is the line of sight from *A*; ship no. 1 is almost completely in view, no. 2 is 'hull down', and no. 3 is indicated only as a puff of smoke on the horizon.

(iii) When you climb higher you can see farther. If there were a lighthouse at *A* (Fig. 16), the line of sight would be *CD* and even ship no. 3 would be mainly visible.

(iv) When an eclipse of the Moon occurs (see p. 46) a shadow of the Earth is cast upon it. That this shadow is curved, though not exactly *proving* anything, does at least suggest that the Earth is round.

(v) The apparent positions of the Sun, Moon and stars change as you travel north and south. This will be referred to again in a later chapter when you have sufficient knowledge to understand it.

At the beginning of the last paragraph it was said that the curvature (i.e. roundness) of the Earth could not be seen. Nor can it by ordinary people, but the navigators of the U.S. balloon *Explorer II* could see it, from a height of 71,000 ft. (13½ miles). One of their photographs, published in 1936 in the *National Geographic Magazine*, shows quite distinctly that the horizon is a curve and not a straight line as in ordinary photographs.

Fig. 17. The dimensions of the Earth

The Earth is not a perfect sphere, like a tennis or billiard ball, but is slightly flattened, more like the 'woods' used for playing bowls. The flattening is at the north and south poles (Fig. 17); the diameter from pole to pole is 7900 miles, and along a line at right angles to that (i.e. at the Equator) it is 7926 miles. Thus you see that the flattening is really not very much, and we usually speak of the Earth as being a sphere 8000 miles in diameter.

CHAPTER III

THE ROTATION OF THE EARTH

ANOTHER fact about the Earth that you know, but may or may not have thought about, is that it rotates upon its axis, i.e. that it is spinning like a top. The line *NS* in Fig. 17 is called the axis, and if you were making a little model of the spinning Earth *NS* is the direction of the needle on which you would make it spin. The obvious effects of the Earth's rotation are (*a*) day and night, and (*b*) the Sun, Moon and stars all appear to make a daily journey around the Earth. These can be illustrated quite easily.

Expt. 9. Place a geography globe some distance from a lamp; the former is the Earth and the latter the Sun. Notice that half the globe is lighted and the other half is in shadow, for light travels in straight lines and cannot get around to the back of it. Now rotate the globe, and all parts become light and dark alternately, just as all the Earth (except polar regions referred to in chapter VIII) have day and night.

Expt. 10. Stand in the middle of the room and notice how various objects on the walls are placed, in front, to your right, to your left. Now slowly turn around in a direction opposite to that of the hands of a watch. Notice how the objects in front of you move away to your right and disappear, while those on the left come in front of you and new ones take their place. If you live in a room which faces south you will have noticed that the Sun is to your left in the morning and to your right in the evening.

Expt. 9 shows you the cause of night and day and No. 10 shows why the Sun, Moon and most of the stars rise in the east, move across the sky, and set in the west. In the far off days when people believed in a flat Earth they supposed that there was a river around it (Fig. 18). A boat carrying a fire sailed around once a day, and in the evening, when it was in the west, it went behind the mountains, which screened the world from the light of the fire until it reappeared in the east in the morning. Later on it became known that the Earth was round, and then the explanation of

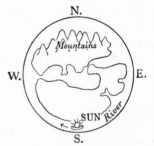

Fig. 18.   The ancient world

night and day was that the Sun moved around the Earth, as indeed it appears to do. However, the Moon and thousands of stars also move across the sky in the same way, and it is much more reasonable to suppose that they are fixed and that the Earth turns around. We know from other scientific experiments that this is the true explanation.

The rotation of the Earth makes the sky seem to rotate, and as the sky rotates from east to west we know that the motion of the Earth must be in the opposite direction, from west to east. Let us make a model of the rotating sky.

Expt. 11. Put water into a round-bottomed flask, sufficient in quantity to fill the round part exactly half way when the flask is corked up and inverted. Hold the flask in the position shown in Fig. 19 and rotate it as shown by the arrows. The water represents the sea, the glass the sky, and the water line around the glass the horizon. Make an ink spot at about *A*, and notice how it rises in the east, moves over the sky, and sets in the west. A spot at *B* behaves similarly,

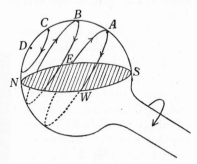

FIG. 19.   Model of the rotating sky

but notice that it remains above the horizon very much longer than *A*. A spot at *C* never sets at all; it just describes a series of circles, at the centre of which is a spot *D*, which does not move.

This experiment reproduces almost exactly the motion of the stars in the sky. Most of them are like *A* and *B*, rising and setting in the same way as the Sun, but others do not set at all. When they sink in the west they curve around near the northern horizon and then climb up again in the east. The point *D*, about which the stars seem to revolve, is called the *north pole of the sky*, and the stars around it, which never set, are the *circumpolar stars*.

In order to photograph stars a time exposure is necessary, i.e. they are too faint to show in a snapshot and the shutter must be left open for a long time, minutes or hours according to circumstances, to give the faint light time enough to make an impression on the film. If such a photograph be taken with an ordinary camera in a fixed position, the stars will not appear as points but as streaks, for they have moved across the sky during the exposure.

Fig. 21 is such a photograph, taken with the camera pointed fairly high towards the northern sky, and it demonstrates very clearly that the stars in this direction do move in circles. One of the trails is brighter than the others and is very close to the centre; this is made by the *Pole Star*, a fairly bright star about which all the others seem to revolve. The photograph shows that it is not quite at the pole of the sky, and therefore not quite at *D* in Expt. 11.

The Pole Star is in the north, and may therefore be used as a guide to direction. Thus you should learn how to recognise it. The brighter stars are named in groups, called constellations, and a well-known group is that called the *Plough*,[1] which it resembles, or the *Great Bear*, which it does not. This is shown in Fig. 20, and on any fine night there should be no difficulty in finding it. The two stars marked *A* and *B* are called 'the pointers', because if you follow the direction of the arrow for

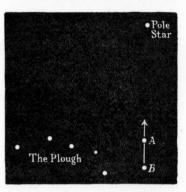

FIG. 20.   To find the Pole Star

about four times the distance from *A* to *B* you will find the Pole Star.

OBSERVATIONS

(1) Notice in the south the position of a bright star with respect to some landmark such as a chimney. In an hour return to the same place and look again.

(2) Find the Plough and the Pole Star.

(3) Notice the direction of the Plough from the Pole Star, and look again several hours later.

The following are for photographers:

(4) Point your camera at the Moon, open the shutter, and leave it for an hour.

[1] In America, 'the Dipper'.

(5) Repeat (4) for bright stars in the south on a night when there is no Moon.

(6) Take a photograph like Fig. 21. Essential conditions are a clear sky, little or no wind, no Moon, no lights near the camera, and as long an exposure as possible.

CHAPTER IV

CONSTELLATIONS

ASTRONOMY is a subject that you cannot do entirely at your desk; you must go out and look at the sky sometimes. It is therefore necessary to be able to find your way about there, and in this chapter we shall study the map of the sky. You will have realised by now that the sky is always changing, both from hour to hour and from day to day, and one map cannot always be right. The two maps given as Figs. 22, 23 are correct only on the dates and at the times mentioned below them, but between them they show all the chief constellations that can be seen during the winter evenings. The middle of the map represents the zenith, which is the point directly overhead, and the circular edge is the horizon. If you hold the map over your head with the word 'north' towards the north you can compare it with the sky.

The constellation groups were named many centuries ago, and this accounts for their peculiar names. In a few cases, such as those of the Lion and the Flying Swan, the shape of the star group does slightly resemble the subject named, while in others the ancient people imagined the outline of some national hero, or of the characters in one of their legends. The following are names taken in this way from the old Greek stories: Cassiopeia, Perseus, Andromeda, Pegasus, Castor, Pollux and Orion.

The brightest stars are said to be of the *first magnitude*; these are marked on the maps as larger dots, and their names are given.

During the autumn evenings the Great Bear (or Plough or Dipper) will be found in the north, rather low in the sky, and above it, in line with the pointers, is the Pole Star. On the opposite side

of the Pole Star from the Great Bear, and therefore nearly over-
head at this time of year, is the constellation called Cassiopeia,
shaped like a large letter **W**. These groups, the Great Bear and
Cassiopeia, are circumpolar and therefore never set in the British
Isles, though stars which are a little farther than these from the
Pole Star do set, and in certain seasons cannot be seen at all.

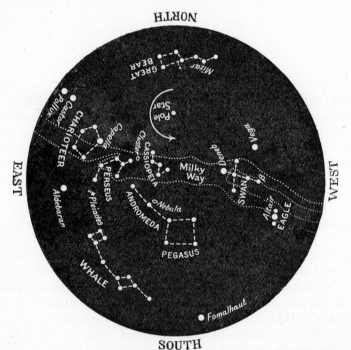

FIG. 22.   The stars of autumn. Oct. 1, 11.0 p.m.; Nov. 1, 9.0 p.m.;
Dec. 1, 7.0 p.m.

Imagine a line from the Pole Star to the western end of Cassiopeia, and follow it on beyond the latter for about an equal
distance. To do this you will have to turn around and face the
south, where you will then see a large and very distinctive square
called Pegasus, well above the horizon. The top left-hand corner
as you are now looking at it really belongs to Andromeda, a line

of three widely spaced stars that make Pegasus look rather like the Plough on a large scale. Return to Cassiopeia. From its eastern end there is a line of stars slightly curved and pointing towards the horizon; this is Perseus, and at the end of it is a pretty little group, like a tiny Plough, called the Pleiades. The Pleiades do not belong to Perseus but, like the first magnitude Aldebaran,

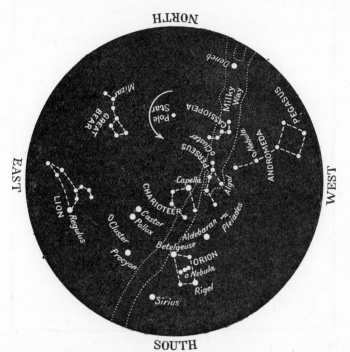

Fig. 23. The stars of winter. Jan. 1, 11.0 p.m.; Feb. 1, 9.0 p.m.; March 1, 7.0 p.m.

to the Bull. Close to Perseus (refer to map) is another first magnitude star, Capella, and still nearer the horizon two more, Castor and Pollux, the Twins. In the western sky, to the right of Pegasus, there is a large cross with a first magnitude star at the head of it. This constellation is the Swan, shown in old maps with its wings outstretched and having the bright star Deneb at its

tail. Below the Swan is the Eagle, a bright star, Altair, with a
fainter one on each side, and a little farther round towards the
north is another first magnitude star, Vega.

After Christmas and in the depth of winter the sky has changed
(Fig. 23). The Eagle and the Swan have disappeared, Pegasus is

FIG. 24. The constellation of Orion

low in the west, while Perseus and Capella are now overhead. The
southern sky is occupied by a magnificent group of stars including
in all no fewer than seven of the first magnitude. The central
figure in this display is Orion, separately illustrated in Fig. 24,
which is based on a very old star atlas and shows the hunter's
figure. To the right and above Orion is Aldebaran, while on the
left are Sirius, the brightest star in the sky, and Procyon. High
in the sky above Procyon are Castor and Pollux. Only in winter

do we have such a fine display of first magnitude stars. The Great Bear is now in the north-east, tail (or handle) downwards, and if the pointers be followed away from the Pole Star you will come to the Lion, a constellation in which Regulus is one more to add to the list of brilliant winter stars.

As spring comes Orion will set, the Lion will take its place in the south, and new constellations will come up to command the summer sky. There is not space in this book to describe them, but some day the reader should obtain another book, such as the author's *Guide to the Sky*, and study the stars of other seasons.

OBSERVATIONS

(7) Find the constellations described in this chapter and study them regularly until you can recognise them at a glance.

(8) Observe how the directions in which the constellations are seen varies from month to month, observing always at the same time in the evening.

(9) If possible obtain a planisphere. This is a revolving star map that can be set for any hour of any day of the year. Compare it with the sky as often as possible. (Messrs Philip have them, obtainable through booksellers, at 4*s*. each and a larger size at 7*s*. 6*d*.)

CHAPTER V

THE ANNUAL MOTION OF THE EARTH

IN CHAPTER I we learnt that the Earth is a sphere, and in chapter II that it rotates on its axis. It also revolves around the Sun; what evidence is there for that statement? Remembering your geography books you will probably say 'seasons', and that is right, but as the seasons are due to something else besides the Earth's annual motion that phenomenon is given a chapter to itself. Some other evidence will be considered here.

Expt. 12. Place a lamp on the table to represent the Sun, and near it a ball or globe for the Earth. Imagine that the walls of the room represent the distant background of stars. Notice the day and night halves of the Earth; stars can be seen only from the night half, and you will be able to see which part of the walls are providing the starry background. Now move the ball slowly around the lamp, and note that the night half is changing direction

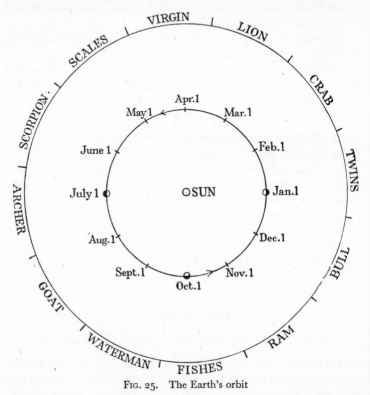

FIG. 25.   The Earth's orbit

all the time and points towards all the walls in turn. This is why we see different constellations at different times of year, and why Figs. 22 and 23 are different.

Examine Fig. 25, which shows the path, or *orbit*, of the Earth around the Sun, together with the directions of certain constellations. On January 1 the night side of the Earth is directed towards the constellations on the right, the Lion, the Crab, the Twins, and the Bull, so they are among the winter constellations. The groups on the left, such as the Archer and the Goat, are in the same direction as the Sun and therefore cannot be seen. On July 1 the conditions are reversed, the winter constellations have disappeared and the Archer and the Goat will be in the summer sky. The circumpolar constellations, such as the Great Bear and Cassiopeia, would be in a direction at right angles to the paper and are visible at any time of year.

Fig. 26 shows the October and January positions enlarged. The point A is in the middle of the night half, so the time there would be midnight; at B it would be about 11 p.m. As the distance from the Earth to the Sun is very much less than that of the stars, the latter would look the same from the Sun as from the Earth, so

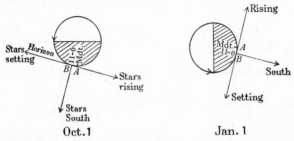

FIG. 26.   October and January positions enlarged from Fig. 25

imagine Fig. 26 (October 1) to be placed in the centre of the circle of constellations in Fig. 25. 'Stars rising' would point between the Twins and the Bull, the latter would be between rising and southing, while the Lion would be out of sight. Compare this statement with Fig. 22, which represents the sky at 11.0 p.m. on October 1. A similar examination of Fig. 26 (January 1) shows that then the Twins and the Bull would be in the south and the Lion rising; look at Fig. 23.

Thus the annual journey of the Earth around the Sun causes the variation in the appearance of the sky.

If the stars could be seen during the daytime we should notice that the Sun moves among them; the reason why we do not see them is simply that their feeble light is overpowered by the very much brighter sunlight. During January (Fig. 25) the Sun is in front of the constellations of the Archer and Goat, in February it

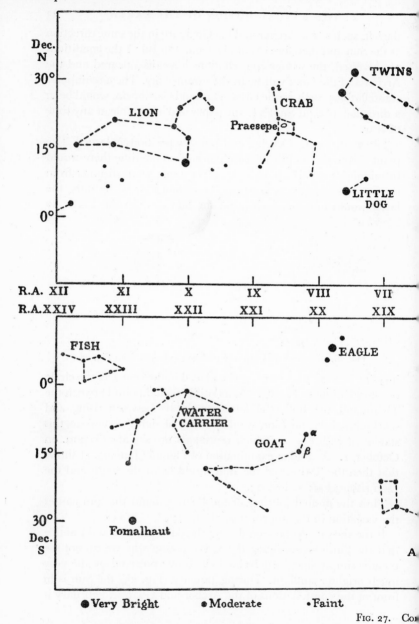

FIG. 27. Cor

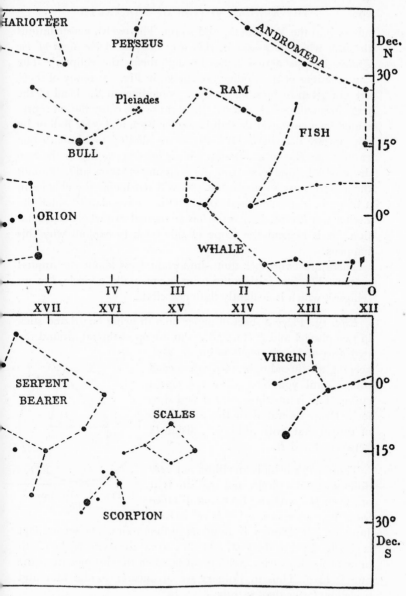

HARIOTEER

PERSEUS

ANDROMEDA

Dec. N

30°

RAM

Pleiades

FISH

15°

BULL

ORION

0°

WHALE

V   IV   III   II   I   O
XVII   XVI   XV   XIV   XIII   XII

VIRGIN

SERPENT
BEARER

0°

SCALES

15°

SCORPION

30°
Dec. S

R.A. and Dec. explained on p. 63

of the Zodiac

moves into the Waterman, and so on. The twelve constellations through which it passes in this way are called *the Signs of the Zodiac*, and its actual path through them 'the ecliptic'. The constellations of the Zodiac are shown in Fig. 27; many of them are not given in Figs. 22 and 23. Sometimes at the head of the page in an almanack you see the appropriate constellation represented by a curious little sign like γ for Ram and 8 for Bull or by appropriate drawings. These signs are allotted to a month one earlier than shown in Fig. 25. This is because each year the Sun drops a bit behind its starting point among the stars, and although the amounts are so small that you will not notice the change in a lifetime, in the hundreds of years that have elapsed since the Zodiac was invented the error has amounted to a whole constellation. It is beyond the scope of this book to explain why this happens.

There are two other quite important facts about the annual motion. The first is that the orbit is not really a circle, but an 'ellipse', which is a slightly flattened circle.

Expt. 13. Place a sheet of paper on a drawing board and stick in two pins, A and B (Fig. 28). Put a loop of thread around the two pins and the pencil point C, and, keeping the thread tight, move the pencil round until you have drawn a closed curve. This is an ellipse and A and B its foci. Draw several with the same loop of thread but with different distances between A and B.

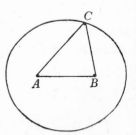

FIG. 28. To draw an ellipse

The Earth's orbit is an ellipse not very different from a circle, and the Sun is at one focus. Thus during a year our distance from the Sun varies, being least early in January. The average is about 93 million miles. The second fact is closely allied to the first, and that is that the speed of the Earth in its orbit also varies and is greatest when the distance from the Sun is least. Thus in January we are nearer to the Sun and travelling faster than in July.

OBSERVATIONS

Observation 8 should be made if not already started.

(10) Choose some fairly bright star in the south-west and observe how far it is from the Sun at sunset. Observe it regularly for several weeks, and notice which way the Sun appears to move among the stars.

(11) Try to find objects which will serve to give you a fixed direction, such as the tops of two posts or the edges of two buildings. Find the exact time (with a watch that is *right*, of course) at which a star crosses this line, and repeat the observation on several nights. Does it get earlier or later? What is the daily change in minutes?

CHAPTER VI

TIME

THE rotation of the Earth gives us a natural clock and the day is our unit of time. Suppose that the place where we live is on that part of the Earth turned towards the Sun; the motion of the Earth gradually turns it away from the Sun until one day later it is towards the Sun again. Thus a day is the interval of time between two successive 'turnings towards the Sun', and that interval is divided up into hours, minutes and seconds.

The expression 'turnings towards the Sun' is both awkward and inaccurate, and we must find a better way of saying what we mean. A line along the ground from north to south is sometimes called the *meridian*. When an astronomer uses the word he does not mean that line only, but also an imaginary line in the sky starting on the south point of the horizon, passing through zenith (overhead), and then going down to the north point. The meridian is really a plane, which is a flat surface including the north–south lines both on the ground and in the sky. To represent a meridian on a model of the Earth we should draw a line from the north pole to the south through the place with which we were concerned. To show what the astronomer means by the meridian we should have to cut the sphere into two halves along the line just drawn, and then stick them together again with a sheet of cardboard between; the plane of the cardboard would represent the astronomer's meridian. Now in the last paragraph we spoke of a place being turned towards the sun; it is more exact to say that at that place the sun is on the meridian. The day, therefore, is the interval of time between two successive crossings of the meridian. The crossing of the meridian is called a transit, and the instant at which it occurs is called midday or noon.

How could you find the meridian experimentally? North can be found from the Pole Star, but that has to be done at night and the Pole Star is not visible from all parts of the world. North can also be found by a compass, but this method is not accurate, as there are only a comparatively few places in the world at which the compass points exactly north. South can be found by observing the Sun at midday, or at any other time by suitable use of a watch,[1] but these methods demand a knowledge of time usually not available, and incidentally it is for the purpose of correcting clocks that the meridian is

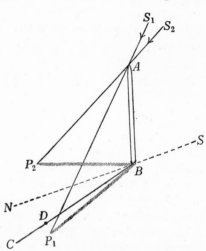

FIG. 29. Finding the meridian

needed. A repetition of Expt. 11, with a mark at a point like A in Fig. 19 for the Sun, will show that when the Sun is on the meridian it is higher above the horizon than at any other time. Thus south can be found by seeing when the Sun is highest and shadows are shortest; the difficulty is that you would not know that it was highest until it had started coming down again, and then it would have passed the meridian. A method of getting over all these difficulties is as follows (Fig. 29):

AB is a vertical post in the ground and BC a string tied to the

[1] Approximately.

bottom. Sometime before noon, say at about 11.0 o'clock, the Sun is in the direction S_1 and the shadow will be directly behind the post, at BP_1, because light travels in straight lines. Mark with a peg, or by other means, the position of P_1, and tie a knot D in the string so that BD is exactly as long as the shadow. Until noon the shadow will shorten as the Sun gets higher, and then it will lengthen again. Use the knotted string to find when the shadow is the same length as it was before, and mark its direction BP_2. The direction is different, of course, because the Sun is moving across the sky all the time. Now the shadow BP_1 must have been formed just as long before noon as BP_2 was after it, and therefore the shadow at noon would be just half way between them. The meridian is the line NBS bisecting the angle P_1BP_2.

When the meridian is known we have a means of measuring time. The interval between successive transits of the Sun over the meridian is a *solar day* and an instrument such as a *sundial*, which divides up this interval into hours, is said to keep *solar time*. The appearance of a sundial is familiar to most people; the edge of the gnomon which casts the shadow is parallel to the axis of the Earth, so that the shadow moves equal angles in equal times, but the method of graduating the horizontal dial, on which the angles are not equal, is rather beyond the scope of this book. The day can also be measured by watching the transits of a star; this kind of day is called a *sidereal day*, and astronomers use clocks that are regulated in this way and keep *sidereal time*. These two methods of measuring a day are given different names because they do not give quite the same results; solar and sidereal clocks do not agree.

If a star be on the meridian at a certain time, it will be so again when the Earth has rotated exactly once, for the stars are so far away that the motion of the Earth around the Sun makes no appreciable difference to their directions. Thus one rotation takes one sidereal day. Suppose that the Earth is at A (Fig. 30), and that the Sun is on the meridian at a place a. When it has rotated once it will have moved on to B, and you can see that the Sun will not be on the meridian at a until the Earth has rotated through the extra bit ab, which takes about 4 minutes. Thus the solar day is 4 minutes longer than the sidereal, and astronomers' sidereal clocks gain 4 minutes a day on ordinary clocks. This is connected with the changes in the constellations that we see week by week, for what was the result of Observation 10 in the last chapter? If a star is on the meridian at 7.0 p.m. to-night it will be there again 24 sidereal hours later, which will be 6.56 p.m. to-morrow. The next transit will be at 6.52, the next at 6.48, and so on, until after a year the daily 4 minutes will have added up to 24 hours and the star will once more be in its place at 7.0 p.m.[1]

You must have heard the expression 'Greenwich Mean Time', and you may have noticed that a sundial is nearly always wrong. The Earth's orbit is not a perfect circle, and as a result the Earth's motion is not perfectly regular, which means that the distance *AB* in Fig. 30 is not quite the same every day. Thus the extra rotation *ab* varies a bit and solar days are not always the same length. Good clocks give us days that are all the same, an average of the solar days, and are said to give *mean time*. The difference between solar and mean time is called the 'equation of time',[2]

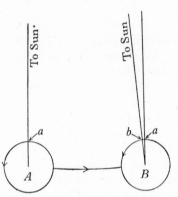

FIG. 30. Sidereal and solar days

and this is given a plus sign when the sundial is ahead of the clock, as it usually is in summer and winter, and minus when behind in the other two seasons. The standard mean-time clock, from which all other clocks in Britain are corrected, is kept at Greenwich Observatory, London, from which the wireless time signals are sent out. The six pips are a second apart, the last one being the exact time.

Summer time is just Greenwich Mean Time plus 1 hour, clocks being put forward an hour in April and put back one in October. Most people are in the habit of not getting up until long after

[1] Approximately. Astronomers would not agree with this, but it is quite near enough for present readers.

[2] The equation of time is also affected by the Earth's tilt, dealt with in chapter VIII, which complicates its variation throughout the year.

daybreak, and not going to bed until long after dark. During the Great War of 1914–18 it was realised that a great deal of gas and electricity would be saved if people would get up earlier and use the morning daylight. Thus 'daylight saving' was introduced, and by putting the clocks on an hour early rising took place without affecting such habits as school at 9.0 a.m. and afternoon tea at 4.30 p.m. The extra hour of daylight in the evening, when work was over, proved so popular that summer time is now a regular custom. When the second Great War came it was decided to leave summer time in force throughout the year and to put on yet another hour from April to August or September.

The 24-*hour clock* is a means of avoiding the use of a.m. and p.m., which we must use if the numbers 1 to 12 come twice a day. The day begins at midnight and the twelve hours to midday are numbered as usual, but instead of beginning again at 1.0 it is called 13.00 and the hours go on to 24. Thus 16.30 hours is what we should call 4.30 p.m. This time system is used by scientists, by the Navy, Army and Air Force, and in Continental railway time-tables. The B.B.C. used it for a few weeks in the *Radio Times* just to see how the British public liked the idea; they didn't—16.30 hours sounds such an awkward time for tea!

Now we must consider the much larger unit of time, the year, in which the Earth makes its annual journey around the Sun. Unfortunately this journey does not take an exact number of days, and is usually quoted as $365\frac{1}{4}$.

Suppose that the year begins when the Earth is at A (Fig. 31); 365 days later it would be at B, not quite where it started, though another day's journey would take it past A. After another 365 days the Earth would be at C, in three such years at D, and in four at E. The distance from E to A is one day's journey, so by allowing one extra day in the fourth year the Earth again starts a new year at A. Thus we have three years of 365 days and then one, *Leap Year*, of 366, the extra day being February 29. This arrange-

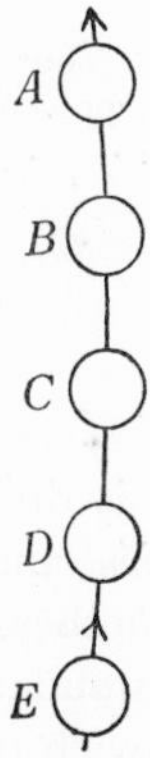

Fig. 31.   Positions of the Earth at intervals of 365 days

ment was introduced by Julius Caesar in 46 B.C., but is not quite correct, because the extra day carries the Earth a little beyond *A*. To correct for this a leap year is left out every 100 years, but this correction is a bit too much, so an extra one has to be put in every 400 years. The result of these modifications, made by Pope Gregory in 1582, is that the century years are not leap years unless the first two figures will divide by four, i.e. 1900 was not a leap year but 2000 will be. The next leap year will be 1948, 48 being divisible by 4.

OBSERVATIONS

(12) Determine the position of the meridian by the method described in this chapter.

(13) If a sundial is available, compare it with correct wireless clock time on as many days as possible for a year. Tabulate your results:

Date	Clock	Sundial	Difference

Note. Your longitude (next chapter) will cause a constant difference, but will not prevent you from detecting the variable change due to the equation of time.

CHAPTER VII

POSITION UPON THE EARTH

TIME is different in different places. Every place has its own *local time*, though as a matter of convenience all the places near one another, as in a small country like ours, set their clocks to some standard time. It was mentioned in the last chapter that Greenwich time is used all over the British Isles. In Fig. 32 the meridian at London (Greenwich) is shown pointing to the Sun, and it is midday there. The New York meridian will not point to the Sun for some hours, and actually the local time there is about 7.0 a.m., while it will be night at Tokyo and Vancouver. You can estimate for yourself what their local time will be, *M* in the diagram being the place where it is midnight. Travellers must alter their clocks as they go. On an Atlantic liner steaming west the clocks are put back an hour every day, while if you travel across Europe by train you put your watch on an hour when you pass from Belgium, which keeps Greenwich Mean Time (G.M.T.), into Germany, which keeps Central European Time (C.E.T.). The world is divided into 'time zones' differing by 1 hour, each zone being 15° wide with some modifications for political boundaries.

This change in time is utilised to determine the position of a place on the Earth's surface. The angle between the meridian at London and that at New York is 74°, and we say that the *Longitude* of New York is 74° West of Greenwich, or, more simply, Long. 74° W. The Earth turns once around, i.e. 360°, in 24 hours, and in 1 hour it will turn through 15°. Thus if we know the difference in time be-tween the two places we can calculate the angle. We noticed

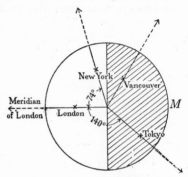

Fig. 32. Longitude

just now that when it is noon (12.0 midday) in London it is about 7.0 a.m. in New York, which is a difference of five hours, and in 5 hours the Earth would turn through 5 × 15 = 75°. At Long. 180° clocks would be 12 hours fast, compared with Greenwich, on one side and 12 hours slow on the other, so when a ship crosses the *Date Line*, which follows this meridian approximately, the date must be changed by one day.

A simple method for finding longitude is as follows: note the correct Greenwich time (wireless time signal) when you mark the shadow BP_1 (Fig. 29), and again for BP_2. Local noon occurs halfway between these times, and if it occurs after Greenwich noon you are west of Greenwich, before it, east.

Example. BP_1, 11.40 a.m.; BP_2, 1.0 p.m. Hence local noon was at 12.20 p.m. G.M.T. As this is behind Greenwich the longitude is west, and if 1 hour corresponds to 15° 20 minutes will mean 5°. The longitude is therefore 5° W. This illustrates the principle of the process; in practice an almanack would have to be used in order to make allowance for the equation of time.[1] This difficulty can be overcome by timing the transit of a star instead of the Sun, for from *Whitaker's Almanack* the times of transit at Greenwich can be worked out and then you can compare this with its time at your own station (see Observation 14).

Longitude alone does not fix your position; it only tells you how far round the Earth you are, measured east or west from Greenwich. You must also know your distance north or south,

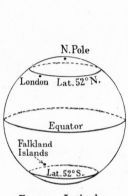

FIG. 33.   Latitude

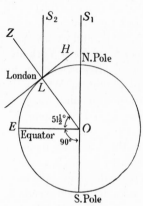

FIG. 34.   The altitude of the Pole Star

and this measurement is *Latitude*. You already know what the poles are. If the Earth could be sliced in two exactly midway between them the line where the surface is cut would be the Equator (Fig. 33). At the centre of the earth the angle between either pole and any point on the Equator is 90° (Fig. 34). The latitude of a place is the angle at the centre of the earth between that place and a point on the Equator exactly north or south of it. London, Lat. $51\frac{1}{2}°$ N., is shown on Figs. 33 and 34, and the Falkland Islands, whose angle is almost the same but south of the

[1] Subtract (or add if −) the equation of time from solar noon to get mean noon. If the E.T. was − 5 minutes in the above example, solar noon at Greenwich would be 12.5 by the clock and the local solar noon obtained only 15 minutes behind.

Equator, Lat. $51\frac{1}{2}°$ S., is also shown in Fig. 33. All places having the same latitude lie on a circle around the Earth, such circles being called *parallels of latitude*.

Expt. 14. Take a geography globe and read this chapter again. Find the places mentioned, and verify as far as you can all the facts stated. It would be helpful if the globe could be illuminated from one side only, preferably by a lamp; it will be midday at points on the globe nearest to the lamp. Notice that whereas meridians of longitude are great circles, all going right around the Earth, the parallels of latitude get smaller and smaller as you get nearer to the poles.

Examine Fig. 34. OS_1 is a line from the centre of the Earth to the Pole Star, and LS_2 from the point we called London to the Pole Star. As the stars are a very great distance away (the paper would have to be about 200,000 miles long to get the Pole Star in the drawing) these two lines may be considered parallel. LH is the horizon at L and is at right angles to LO. Now it can be proved[1] that angle S_2LH is equal to angle LOE, and therefore a simple way of finding latitude is to measure the altitude, or angle above the horizon, of the Pole Star (see Observation 15). Here again an almanack would be wanted for an accurate result, as a correction is necessary for the fact that the Pole Star is not quite above the north pole (see p. 16).

Navigation. If the latitude and longitude of a place are known, its position upon the Earth's surface is fixed. Determinations of these things must be made regularly by the navigator of a ship, the Sun being the body usually observed for the whole process. The navigator's methods are different from ours, but his requirements are the same; he needs, as we do, three things.

(1) *Correct Greenwich time.* A ship's clock is called a *chronometer*; it is specially designed to keep good time in spite of the motion of the ship, and the testing of these clocks is a part of the work at Greenwich Observatory. Before leaving harbour the captain must

---

[1] Because LS_2 and OS_1 are parallel and LO meets them

$$\angle S_2LZ = \angle LOS_1 \text{ (corresponding)}.$$

Hence the complement of $\angle S_2LZ$ = comp. of $\angle LOS_1$

or $\qquad \angle S_2LH = \angle LOE.$

get the correct time, and that is why at various ports there are time guns, or time balls like the one that drops at 1.0 o'clock at Greenwich (visible in Fig. 60 A). The navigator used to be dependent upon the accurate running of his chronometer throughout the whole voyage, but now he can check it by wireless time signals, several times a day if he wishes. A chronometer is shown in Fig. 36 and it can be seen that it is mounted on pivots attached to a ring which is itself pivoted on an axis at right angles to that of the first pair. It is said to be mounted on gimbals, and it will consequently remain level when the case is tilted.

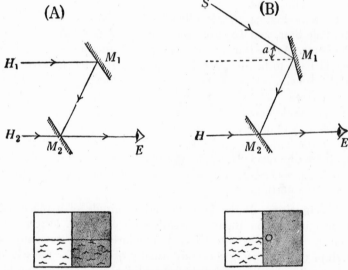

FIG. 35. The principle of the sextant

(2) *An almanack.* The *Nautical Almanack* is published by the government to give mariners all the astronomical information they need, and supplying information for this book is another of the duties of Greenwich Observatory.

(3) *An instrument for measuring angles in the sky.* The *sextant* is the portable instrument used for the purpose and is illustrated diagrammatically in Fig. 35. M_1 can be rotated. M_2 is fixed, and half of it is unsilvered. From the laws of reflection we know that $H_1 M_1$ will be parallel with $H_2 M_2$, and therefore both rays will have

FIG. 36. A ship's chronometer

M_1

FIG. 37. A sextant

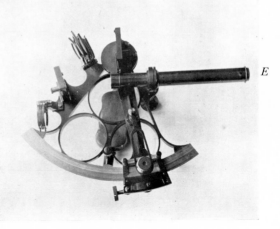

E

come from the same point in the distance. When the instrument is pointed to the horizon the eye will see it twice, through the clear part of M_2, and by reflection in the mirrors, as shown in the inset under the diagram (though without a sharp boundary between the two). When M_1 is rotated it will reflect light from the sky instead of the horizon, and a mariner adjusts his instrument so that he sees the horizon through the clear glass and the Sun by reflection. This is illustrated in B. Now we know from Expt. 2 that the angle a, which is the altitude of the Sun, will be twice the angle through which the mirror has been rotated. Fig. 37 shows an actual instrument, and you can see that M_1 is attached to an arm working over a degree scale, each degree being marked as two so that the scale reading is the altitude. Note also that a small telescope is included in the instrument, and that dark glasses can be inserted in the path of the light to protect the observer's eyes.

Before a sextant observation can be used for calculation, corrections must be applied for several errors: (a) the index error, due to the instrument itself; (b) dip of the horizon, due to the observer not being at sea level; (c) refraction of the atmosphere, explained in chapter 1; (d) parallax, due to the observer not being at the centre of the Earth, which except in the case of the stars makes an appreciable difference to the apparent position of an object; (e) semi-diameter of the Sun or Moon, for while it is usually the lower limb that is observed it is the altitude of the centre that is used in the calculation.

There are several ways of finding the ship's position; the following is a brief outline of what is commonly done. In the early morning the navigator takes the altitude of the Sun and notes the time with a deck watch that can afterwards be compared with the chronometer. The calculation that follows, using the *Nautical Almanack* and specially prepared mathematical tables, gives a position line drawn upon the chart; the position of the ship is somewhere on this line. A few hours later, possibly at noon, another sight is taken, yielding another position line. The ship has, of course, moved in the interval, but knowing her course and speed the run can be determined, and the first position line transferred to the place it would have occupied had it referred to the same instant as the second. The intersection of the two lines is the observed position of the ship. Another method is to take sights of two or more bodies simultaneously or, more correctly, in quick succession, each of which yields a position line. As these bodies are probably stars or planets the observation is made at dawn or dusk, when the horizon is visible as well as the bodies concerned.

The airman must deal with similar problems, but owing to the high speed of aircraft the result is wanted very quickly, although as a rule it need not be quite so accurate. The mathematical work is therefore simplified and he is provided with a special *Air Almanack*. Further, to an observer moving rapidly at a great height, possibly above the clouds, an ordinary sextant would be of little use, so the airman is provided with a special pattern called the *bubble sextant* which enables altitudes to be taken without the use of a visible horizon.

OBSERVATIONS

(14) Erect two *thin*, preferably pointed, posts on the meridian found in Observation 12. Find the time, correct to the nearest second, at which some prominent star crosses the line of the posts. The watch must be corrected by wireless time signal. From *Whitaker's Almanack* (in which the necessary calculation is explained) find the time of its transit at Greenwich. Deduce your longitude from the difference in time (see p. 33). You can now correct for longitude in Observation 13.

(15) Make an apparatus as illustrated in Fig. 38. Observe the Pole Star through the slots, and get a companion to read the angle marked by the plumb line. This is the altitude of the star, and hence your latitude (p. 35).

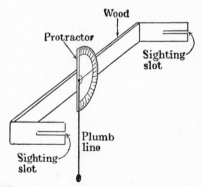

FIG. 38.   An apparatus to measure altitude

CHAPTER VIII
THE SEASONS

CHAPTER v was devoted to the Earth's annual journey around the Sun, but one of the effects of this motion was deferred for the time being. The seasons are due to the combined effect of the annual motion and the tilt of the Earth's axis. Speaking of the tilt of the axis at once raises the question 'tilted to what', for such terms as 'horizontal' and 'up' and 'down' do not apply in Astronomy; 'up' in England is the same direction as 'down' in New Zealand. The Sun appears to move around the sky, through the Signs of the Zodiac, in a year, and its path among the stars is called the ecliptic (p. 26). This path lies in the plane of the Earth's orbit, which is therefore called the *plane of the ecliptic*. By the plane of the orbit we mean the flat surface upon which it might be drawn, and thus when we do draw the Sun, the Earth, and the orbit of the Earth the paper or blackboard represents the plane of the ecliptic. So far we have assumed that the axis on which the Earth spins is at right angles to this plane, and it is shown as such in Figs. 26, 30 and 32. In reality it is not quite like this, and the axis is tilted out of the perpendicular by $23\frac{1}{2}°$ (Fig. 40). This angle is called the 'obliquity of the ecliptic'. You probably know that a spinning object, like a top, is very reluctant to change the direction of its spin. This principle is used in the gyroscope, which is really a heavy high-speed top. By its use a torpedo is guided; there was once a tramway in which the cars stood upright on one rail; the Italian liner *Conte de Savoia* was originally stabilised from rolling; the Earth's axis always points to the same place, the Pole Star. These are all examples of the spinning top. However, we must

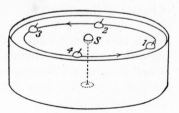

FIG. 39   To illustrate the plnae of the ecliptic.

concentrate our thoughts on the Earth's tilt, and the following little demonstration should make the matter clearer (Fig. 39).

S is a floating ball, anchored in the middle of the bowl by being tied to a weight lying on the bottom. This represents the Sun; the Earth is represented by another ball having pushed through it a needle to show the direction of the axis. The 'Earth' is floated with the axis tilted a little towards one wall of the room, and then moved around the 'Sun', the axis being pointed at the same wall all the time. Here is a very good representation of the motion of the Earth, for the surface of the water is the plane of the ecliptic, and we can see clearly what we mean by speaking of the tilt of the axis, and by saying that the tilt is always in the same direction.

That this tilt affects the condition under which we live can be shown by the experiment illustrated diagrammatically in Fig. 40.

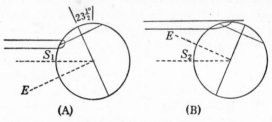

FIG. 40.   The Earth's tilt

A narrow beam of light is directed horizontally on to a geography globe so that the British Isles lie in the centre of the illuminated part. In A the axis is set towards, and in B away from, the source of light, and the difference is obvious. The beam of light is spread over a much larger area in B than in A, with the result that the British Isles are less brightly lighted. Applying this to the real Earth we see that the radiation from the Sun, which includes heat as well as light, is less intense when the axis is turned away from the Sun than when turned towards it. Hence our seasons of winter and summer.

Expt. 15. The room must be darkened except for one lamp, at eye level, and the apparatus required is just an old tennis ball with a steel knitting needle pushed through it to represent the axis.

(i) Hold the 'Earth' in the summer position (Fig. 39, no. 3), i.e. with the axis slightly tilted towards the 'Sun', and slowly rotate it. Notice that the upper half of the ball, representing the northern hemisphere of the Earth, is more than half illuminated, and that the area near the north pole is light all the time (Fig. 41A).

(ii) Hold the 'Earth' in the winter position (Fig. 39, no. 1), and notice that the northern hemisphere is less than half illuminated and that near the north pole it is in darkness all the time (Fig. 41 B).

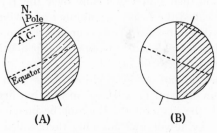

FIG. 41. The seasons

(iii) Now hold it in a position represented by 2 or 4 in Fig. 39, the axis being tilted across the direction of the light, neither towards it nor away from it. Notice that the northern hemisphere is just half illuminated, and that every part of the Earth receives light for half a rotation.

This experiment illustrates several important things. In the first place, in summer we receive more than our fair share of light, and therefore the days are longer than the nights. In winter the nights are longer than the days. Next, there is a region near the north pole, bounded by the *Arctic Circle*, where in summer it is day all the time. This means that the Sun never sets, but just skirts the northern horizon instead, and many of the pleasure cruises to the Norwegian Fjords visit the North Cape to see the 'midnight sun'. In winter, of course, the Arctic regions have perpetual night. At the time illustrated in Fig. 40 A, when the Earth is tilted directly towards the Sun, the Sun will appear overhead at places having the latitude of S_1. This line is called the *Tropic of Cancer* because at the time when the condition occurs the Sun is in the Zodiacal sign of the Crab. S_2 in the winter diagram lies on the *Tropic of Capricorn* (Capricornus: the sea goat). By 'the tropics' we usually mean the zone of the Earth between these two lines and in which the Sun can be seen exactly overhead at some time during the year.

Finally, in spring and autumn there is a time when the whole world is receiving equal treatment in the matter of light, and the day is everywhere the same length, about equal to that of the night. The Sun then shines vertically over the Equator. These positions are called the *Equinoxes*; summer and winter are called *Solstices*. The dates at which they occur are: Equinoxes (2 and 4 in Fig. 39) March 21–22 and September 23–24; Solstices (1 and 3) June 21–22 and December 22.

How do seasons affect the position of the Sun in the sky? A modification of Expt. 11 will explain this, but first refer again to Fig. 40. If the plane of the Equator be imagined to extend right into the sky, it will trace out a line there called the 'celestial equator'. You can find out roughly where it is by bending a piece of wire into a right angle, pointing one leg at the Pole Star, and twisting that leg; the other leg will move around, pointing all the time towards the celestial equator. Notice the important fact that the angle between a point on the Equator and the pole is always 90°. Fig. 40 A shows us that in summer the Sun is above the Equator, and in winter, B, below it; we may reasonably suppose that between these two positions the Sun will be on it, and this is, in fact, the real definition of the Equinoxes, the instant when the Sun is on the Equator.

Expt. 16. Take the flask filled for Expt. 11, hold it with the neck vertically downwards, and draw a chalk line around the water level to represent the Equator, everywhere 90° from the pole D (Fig. 19). Now tilt the flask into its usual position.

(i) Summer: Put a mark for the Sun a little above the Equator and rotate the flask. Notice how it rises N. of E., passes right over the top, i.e. high in the sky, sets N. of W., and is above the horizon longer than below it.

(ii) Equinoxes: Put the mark on the Equator. The Sun rises exactly E., does not pass over as high as in (i), sets exactly W., and is above the horizon for exactly half a rotation.

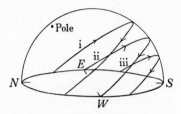

(iii) Winter: Put the mark below the Equator. The Sun rises S. of E., never gets far above the horizon, sets S. of W., and gives a day very much shorter than the night.

FIG. 42. The path of the Sun across the sky

These results are illustrated in Fig. 42. Tilt more steeply for a higher latitude and repeat; try also for a lower latitude.

OBSERVATIONS

(16) Use the apparatus used for Observation 14 to find the altitude of the Sun at midday. Use a dark blue, or thickly smoked, glass to protect your eyes from the brightness of the sunlight.

(17) Use a pocket compass to find the exact direction in which the Sun sets. Allow for magnetic variation if you know its value.

These two observations should be repeated once a week for several months in order to test the results of Expt. 16. Tabulate as under:

Date	Altitude of Sun at midday	Sunset	
		Time	Direction

Sunrise can be included with advantage if you get up in time.

CHAPTER IX

THE MOON

THE EARTH does not travel alone; it has an attendant body, or *satellite*, about 240,000 miles away, moving around it once a month and making the Earth a pleasanter place to live in. The Moon is ever changing in appearance and adds beauty to both land and sea.

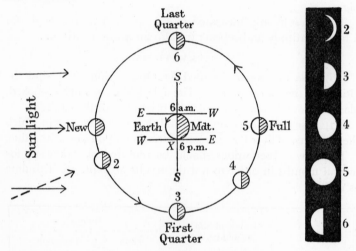

FIG. 43.   The phases of the Moon

When we speak of a 'new' Moon we think of a thin crescent in the western twilight just after the Sun has set. The next day it will still be in the west, but will not reach its former position until rather later in the evening. Each day it will move a little to the east among the stars until at sunset it is only just rising, and later still in the month we do not see it at all in the evening, for it rises too late, but instead we find the 'old' Moon in the west or south-west in the morning. If you are observing at 6.0 p.m., your position in Fig. 43 is at *X*, and you can see how, as the Moon moves on through positions 2, 3, 4 and 5, it is moving towards the east. When at no. 3 it is on the meridian at 6.0 p.m., but at no. 6 it is on the meridian at 6.0 a.m. An interesting exercise is to note its position among the stars regularly (Observation 18) and see how long it takes to come back to the same place among them. This is a little over 27 days, and is therefore the time taken to go once around the Earth. In its passage among the stars it goes in front of some of them, these phenomena being called *occultations*. The observation and timing of the disappearance and reappearance of stars in this way is very important, as the results enable the motion of the Moon to be calculated, and its future positions tabulated in almanacks, with greater accuracy. The time of rotation on its axis is the same as that of its journey around the Earth, so we always see the same side of it.

Fig. 43 also explains the changes in the appearance of the Moon, called its *phases*. The Moon is not a lamp; it is a solid sphere about 2000 miles in diameter; one half being lighted by the Sun.

Expt. 17. The apparatus and lighting should be the same as in Expt. 15. Face the light, hold the ball at arm's length, and turn slowly round towards your left hand until you are again facing the light. Notice the changes that you can see in the amount of ball lighted and compare them with Fig. 43.

Position no. 1 is what is really meant by 'new' Moon; it is in the same direction as the Sun, with the dark side towards the Earth, and is therefore invisible. In no. 2 most of the portion towards us is dark, but there is just a narrow crescent of light on the western side. When it reaches no. 3, a position called 'first quarter' because it is a quarter of the way around, we see half of it, and at 4, rather more than half and say that it is 'gibbous'. No. 5 shows 'full' Moon, with the whole of the lighted half towards us. So far we have considered the 'waxing' (increasing) phases, with the western side illuminated; now we come to the 'waning' (decreasing) phases, which are similar, but with the eastern side showing, like no. 6. Remember that the waxing Moon faces the sunset and the waning Moon the sunrise; artists forget this sometimes and draw the Moon in a perhaps pleasing but quite impossible position.

As seen from the Moon the Earth would have phases, and just as we enjoy moonlight the imaginary inhabitants of the Moon would receive earthlight. Sometimes when the Moon is young and is but a thin crescent the remainder can be seen, looking a dim coppery colour. The bright crescent is illuminated by direct sunlight, and the remainder by light reflected by the Earth, usually called 'earthshine'.

When the Moon has made one revolution around the Earth we should expect new Moon again. The Earth, however, has travelled on in its orbit, and the direction from which the sunlight now comes is shown in Fig. 43 by the dotted arrow. Thus no. 2

FIG. 44.   An eclipse of the Moon

is the new Moon position, and we have to wait for an extra two days for the Moon to get there. Thus a month measured by the Moon's phases is $29\frac{1}{2}$ days, instead of the $27\frac{1}{3}$ given by its motion among the stars.

In Expt. 17 you may have found that the ball representing the Moon went exactly between you (the Earth) and the lamp (the Sun), and in another position entered the shadow of your head. Phenomena of this kind do happen sometimes and are called *eclipses*. They can be demonstrated quite simply by moving a tennis ball around a geography globe placed several feet from a lamp. The shadow of the ball passes over the globe; this is an eclipse of the Sun, or Solar Eclipse, and it causes a short period of darkness over that part of the Earth upon which it falls. The ball moves into the shadow of the globe; this is an eclipse of the Moon, or Lunar Eclipse, when a curved shadow is seen to move over the face of our satellite, making it grow dim for a time (Fig. 44). In the eclipse illustrated (1932) the Moon did not move straight through the middle of the shadow.

Eclipses do not happen every month because the orbit of the Moon is not quite in the plane of the ecliptic (Fig. 39). It is near enough to that plane for the Moon to pass through the constellations of the Zodiac, but tilted at a sufficiently large angle[1] to make eclipses infrequent. In Fig. 45 new Moon

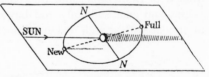

FIG. 45.   Why eclipses are infrequent

occurs at a time when the shadow of the Moon is south of the Earth, and full Moon when our satellite is north of the Earth's shadow. Eclipses occur only if the Moon is new or full when close to the points marked N, where its orbit cuts the plane of the ecliptic.

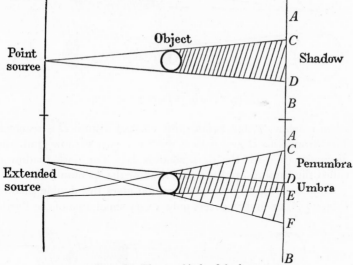

FIG. 46.   The two kinds of shadow

You have probably noticed the difference between shadows due to a window (*not* direct sunshine) and those due to a lamp; the former are hazy and ill-defined, while the latter have sharp edges. The difference is due to the size of the sources of light (Fig. 46).

[1] About 5°.

Suppose that the light is coming from a point source and you are moving from A to B. Until you reach C the source is visible, from C to D invisible, and then visible again. The shadow CD begins and ends suddenly and has sharp edges. Repeating the argument with a wide extended source, it begins to be hidden

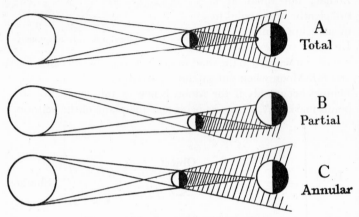

A
Total

B
Partial

C
Annular

FIG. 47. The three types of solar eclipse

when reaching C, but is not quite concealed until D is reached. Thus from C to D and E to F there is a part shadow gradually deepening towards the dark shadow DE. The dark shadow is called the 'umbra' and the part shadow the 'penumbra'.

The Sun is an extended source and gives this double shadow (Fig. 47). The umbra covers only a very small area of the Earth,

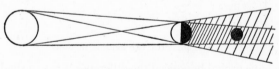

FIG. 48. A lunar eclipse

and therefore a total eclipse, in which the Sun is quite covered, is rather a rare occurrence in any one place. The last such eclipse in Britain was in 1927, over a belt about 30 miles wide, stretching approximately from North Wales to Hartlepool, and the next will be in 1999 in Cornwall. The general appearance of a total eclipse is seen in the frontispiece; note the darkened sky and the planet Venus showing above and to the right of the eclipsed Sun. Places in the penumbra see the Sun's disc only partly covered, as is the case in a partial eclipse (Fig. 47 B), when the umbra misses the Earth altogether.

The orbit of the Moon, like that of the Earth, is an ellipse; thus the distance and apparent size vary a little. If what would have been a total eclipse occurs when the Moon is at its greatest distance from the Earth, the umbra does not quite reach the Earth's surface (Fig. 47 C). This is called an annular eclipse, as when the Moon is exactly in front of the Sun it is too small to cover it and leaves a rim of light all around.

A lunar eclipse is illustrated in Fig. 48. The penumbra in this case is of little importance, for it is the umbra shadow that we notice crossing the Moon's surface, and the eclipse is said to be total if the Moon enters it completely.

One other point before we close this chapter. The Moon moves around the Earth in an ellipse, and the Earth is itself moving around the Sun in another ellipse. The Moon is therefore going around the Sun in an orbit lying close to that of the Earth but intersecting it. Fig. 49 shows a part of it, and the point to notice is that both orbits are concave towards the sun. The diagrams in some atlases are incorrect in this respect.

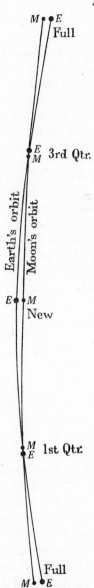

FIG. 49. The Moon's orbit

Observations

(18) Compare Fig. 27 with the stars near the Moon, identify the constellation in which it is situated, and decide on the exact spot on Fig. 27 at which the Moon should be inserted. Make a tracing of that part of Fig. 27 concerned and insert the position of the Moon, with date, and phase if full or first quarter. Repeat this as often as possible for not less than five weeks, and so find the path taken by the Moon among the stars. From it find out the interval of time (*a*) for a full circuit of the sky, and (*b*) for a full cycle of phases, i.e. full to full or first quarter to first quarter.

(19) If a small telescope (or good binoculars) is available, find out from *Whitaker's Almanack* the date and time of an occultation. Start watching a quarter of an hour before the predicted time, and try to observe the exact moment, to the nearest second with a correct watch, at which it does occur. The prediction is for Greenwich, and is slightly different in other parts of the British Isles.

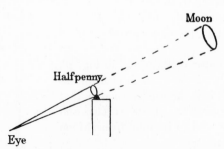

Fig. 50.   Estimating the size of the Moon

(20) Stick a halfpenny into a blob of plasticine on top of a wall (Fig. 50) and then stand in such a position that the Moon is just behind it and fits it exactly. Get a friend to measure the distance (about 10 feet) from your eye to the halfpenny. By the property of similar triangles

$$\frac{\text{Diameter of Moon in miles}}{\text{Distance of Moon in miles}} = \frac{\text{Diameter of coin in inches}}{\text{Distance of coin in inches}}.$$

If the distance of the Moon is 240,000 miles, what is its diameter?

CHAPTER X

THE STORY OF THE PLANETS

CHAPTER IV was devoted to the Constellations, and we saw there that the stars were mapped and named in very ancient times. There were a number of bodies, including the Sun and Moon, which could not be mapped because they were constantly moving among the others, and these were called 'planets', meaning 'wanderers'. The term 'planets' no longer includes the Sun and Moon, but just refers to those which we now know to be revolving around the Sun in the same manner as the Earth. Five of these were known to the ancients. *Mercury* and *Venus* are nearer to the Sun than we are, and hence their orbits lie within that of the Earth. *Mars*, *Jupiter* and *Saturn* have larger orbits than that of the Earth, those of the last two being very much larger. The paths of these planets in the sky lie in the Zodiacal constellations and therefore their orbits, like that of the Moon, are nearly in the plane of the ecliptic. By observing the rates at which the planets moved among the stars the ancients got a very fair idea of their order in distance, but they did not know that they were revolving around the Sun.

The earliest ideas about the Earth were, of course, that it was flat, but by 500 B.C. men had realised that it was actually a sphere, and Eratosthenes of Alexandria (276–196 B.C.) obtained an estimate of its diameter. All the celestial bodies, however, were regarded as being there for our benefit and, like the Moon, revolving around the Earth. The Earth was supposed to be at the centre of a series of crystal (i.e. transparent) spheres. The outermost had the stars attached to it, and it revolved once a day; this did give a reasonable explanation of the observed facts. The Sun and Moon were on smaller spheres revolving, the former a little and the latter considerably, slower than the star sphere; this did not explain the variable rate of movement and these spheres were subsequently modified. The spheres for the planets fell far short of explaining the facts, so let us consider how the planets do move and see what the difficulties were.

Probably the best known planet is Venus, the evening star. When first seen it will be low in the west just after sunset. As the weeks go on it gets higher in the sky and remains above the horizon longer, i.e. it is apparently farther from the Sun. Then its eastward motion stops and, moving more quickly this time, it returns towards the sunset and vanishes. A few weeks later it appears in the eastern sky as a morning star, just before sunrise. It rapidly reaches its maximum distance from the Sun, and then slowly

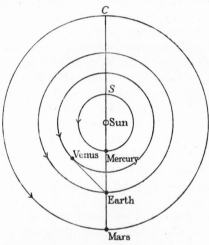

FIG. 51. The innermost planets

returns to it. The motion of Venus can be imitated by moving a tennis ball around a lamp and watching it from a distance. It appears alternately to the left and to the right of the lamp, just as Venus appears to the left of the Sun as an evening star and to the right as a morning star. This experiment also shows that Venus has phases like those of the Moon. Fig. 51 shows the Sun and the four innermost planets. Venus is drawn in the position called 'eastern elongation', because the angle at the Earth between Venus and the Sun is as great as it can be. The corresponding position on the opposite side of the Sun is, of course, 'western elongation'. Mercury shows a similar motion, but being nearer to the Sun never rises very high in the sky and is consequently difficult to see. It is drawn in Fig. 51 in the position called 'inferior conjunction', and when at the point S it is at 'superior conjunction'; planets in conjunction cannot be seen, as they are in the same direction as the Sun. The Earth is all the time moving in its own orbit, but the inner planets move more quickly and overtake the Earth when they are at inferior conjunction. In ancient times Venus was thought to be two planets, Lucifer in the morning and Hesperus in the evening. When the old astronomers realised that these were really the same planet, some, at least, believed that it went around the Sun, although they would not admit that the Earth did so too.

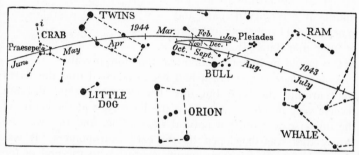

FIG. 52. The path of Mars among the stars

The general motion of the other planets is similar to that of the Sun, a slow journey from west to east in the Zodiac. On careful examination, however, it is found that sometimes they stop, go backwards for a bit, and then go on again, as shown for Mars in Fig. 52. To explain this backward or 'retrograde' motion the ancients gave up the crystal sphere, which would account for the direct motion only, and supposed that the planet moved in a small circle, called an epicycle, whose centre moved around the Earth (Fig. 53). Actually Fig. 51 gives all the explanation needed. Mars, in going around the Sun goes around the Earth also, because the Earth's orbit is entirely within its own; hence the steady west to east motion. The Earth, however, moves more rapidly, and from time to time overtakes Mars, causing the retrograde motion, just as when a train in which you are travelling overtakes another going in the same direction the latter seems to be going backwards. Jupiter and Saturn behave similarly, but as they take so long to go around the Sun (see p. 59) they do not get very far between the periods of retrogression. In Fig. 51 Mars is shown in the position called 'opposition', when the Earth is just passing it and we see it in a direction exactly opposite to that of the Sun. A planet at opposition is in the middle of its retrograde path, and is nearer and brighter than before or after that date. If Mars were

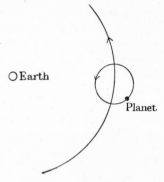

FIG. 53.   An epicycle

at C it would be in conjunction and therefore invisible.

As long ago as 500 B.C. it had been suggested that the Earth went around the Sun, but this 'heliocentric system' did not receive very much support. That the Earth was at the centre of the universe was quite a natural assumption, and this was the theory that was developed by the early astronomers. It was accepted as truth for some twenty centuries, to the end of the period sometimes spoken of as the Dark Ages. This was not simply that there were no intelligent people, but because to question the authority of the old philosophers was just 'not done', and any who dared publicly to dispute their teaching were regarded with extreme disfavour, and sometimes ruthlessly suppressed. With the revival of learning the heliocentric system was again proposed. Copernicus (1473–1543, Poland) studied the matter in detail, and in the year of his death published the book that was to set men thinking on the right lines.

The next development was the work of Tycho Brahé (1546–1601, Denmark). He was interested in the Copernican system, though he rather disapproved of it. His life's work was the making and recording of a large number of observations of the heavenly bodies, to facilitate calculations of their motion. He was very well-to-do when he started this work, but he lost favour with the Court, left Denmark, and died, with the work yet unpublished, in rather poor circumstances in Bohemia. His tables were completed and published by his assistant, Kepler (1571–1630, Germany).

The work of Tycho proved of very great importance; Copernicus gave the idea, Tycho obtained the necessary facts, and then Kepler solved the problem. Applying mathematics to the results of his former chief he deduced three *laws of planetary motion*.[1] The first was that the Earth and other planets moved not in circles but in ellipses, and the second showed how the speed of a planet in its orbit varies. A planet moves most quickly when in that part of the ellipse nearest to the Sun. These two laws cleared one of the difficulties which had given rise to the epicycles previously used, and then the third one gave a numerical relation between the distances of planets from the Sun and their times of revolution around the Sun.

The problem had now been solved, but it remained for Galileo (1564–1642, Italy) to prove that the solution was correct. Galileo heard of the invention of an optical instrument for viewing distant objects, and after making inquiries about it he made one himself. Early in 1610 he started to use it on the heavenly bodies and made many momentous discoveries. Only two will be mentioned here. First, he found that Jupiter had four satellites moving around the planet in just the same way as the planets were said to move around the Sun; this showed that Kepler's system could exist in fact as well as on paper. Secondly, in the experiment with the tennis ball we saw that Venus showed phases; Copernicus foretold them, and Galileo saw them. Thus the telescope provided confirmation of the correctness of the theory, and the era of modern astronomy had begun.

[1] (1) Every planet moves in an ellipse, with the Sun in one of the foci.
(2) The straight line drawn from the centre of the Sun to the centre of the planet sweeps out equal areas in equal times.
(3) The squares of the periodic times of the several planets are proportional to the cubes of their mean distances from the Sun.

OBSERVATION

(21) Very carefully compare Fig. 27 with the sky. If you can find a star in the sky but not on the map, it is probably a planet. Mark its position and date on a tracing, as you did in Observation 18. Watch it regularly, say once a week, to see if you can detect its motion, and insert it on your tracing for as many weeks or months as it remains visible.

CHAPTER XI

THE LAW OF GRAVITATION

THE next step forward was made by a very famous Englishman, Sir Isaac Newton (1642–1727). Kepler had shown how the planets moved; Newton's problem was 'Why should they move like this?' It is common experience that an unsupported body falls to the ground as if the Earth were pulling it down, and Newton's idea was that the same force of gravitation held the Sun and its planets together as one family in space. The fall of an apple in his garden is said to have given him his idea and, whether it did or not, pieces of the apple tree that grew in his garden have been carefully preserved. The result of applying gravitation to Kepler's laws was the great *law of gravitation*, which you will find described in more detail in text books on Mechanics. The substance of the law is that the force between any two bodies is proportional to the two masses multiplied together, and inversely proportional to the square of the distance between them. The first part means, of course, that the greater the body the greater the force. If you fall out of an aeroplane there is an attraction between you and it, but as the Earth is so very much more massive than the aeroplane the former wins the resulting tug-of-war and you fall. In fact the Earth is so much greater than anything on the Earth that its own gravitational pull is the only one of which we are conscious. The second part of the law means that the greater the distance between two bodies the very much less the force between them; double the distance—one-quarter of the force; treble the distance—one-ninth of the force. There are a number of inverse square laws like this, and another will turn up in chapter xv of this book. The law of gravitation was a very powerful weapon in the hands of the astronomer, as it enabled calculations on the heavenly bodies to be made with an accuracy never before obtained, and several examples will be noticed as you go on reading.

The Moon is smaller than the Earth and very much farther away: this makes its gravitational force at the Earth's surface weak, but nevertheless it is strong enough to cause the *tides*. The water on the hemisphere towards the Moon is a little nearer to that body than is the Earth beneath, the Moon's gravitation is a little stronger, and the water is drawn slightly towards the Moon

Fig. 54. The cause of the tides

giving high-tide at *A* (Fig. 54). The Earth is nearer to the Moon than the water at *B* and therefore moves away slightly, giving high-tide at *B* also. Water flows from *C* and *D*, causing low-tides there, to provide for the heaping up at *A* and *B*. As the Earth rotates the positions of *A* and *B* are maintained by the Moon, and thus we have our two tides a day. Needless to say the theory of the tides is a complicated one, and the above is only a very simple outline of it. The Sun also tends to cause tides, and at times of full or new Moon the solar and lunar tides would both be at *A* and *B*; these tides are therefore higher than the average and are called 'spring tides'. At first or last quarter the solar tides should be at *C* and *D*, but as the water cannot do two things at once it has to make a compromise between them. The tidal force of the Moon is the greater of the two, so the tides still occur at *A* and *B*, but are not so high and are called 'neap tides'.

The law of gravitation has led to the discovery of more planets, though the first addition to the six already discovered was an observational one. William Herschel, afterwards Sir William (1738–1822), was a German by birth, but had come to England and settled in Bath as a musician. Astronomy was his hobby and he made his own telescopes; eventually he became one of the great professional astronomers. He was observing one night at his home when he saw a star that was of rather unusual appearance. He watched it regularly and found it to be moving, and thus in 1781 the planet *Uranus* was discovered.

The motion of this planet in its orbit, which is outside that of Saturn, provided material for further mathematical work, and it was found that its behaviour was not quite as it should be. Two mathematicians, Adams in England and Leverrier in France, quite independently tackled the problem of these variations and in 1846 reached almost identical conclusions. They not only found that Uranus was being disturbed by the gravitation of another planet farther out still, but stated just where the new planet should be situated. Quite near to its predicted place *Neptune* was seen for the first time by an astronomer at the Berlin Observatory. This discovery was a triumph for the gravitational work founded by Newton.

History is said to repeat itself, and in this matter of planetary discovery it did. Further study of the motion of Uranus, including the effect of Neptune, was made by Percival Lowell (America; died 1917) who deduced the presence of yet another distant planet. The search at the time was unsuccessful, but in 1930 *Pluto* was found on photographs taken at the Lowell Observatory by another American, C. W. Tombaugh. Thus there are nine major planets now known, and a list, with numerical information, completes this chapter. One cannot leave the story of the planets, however, without remarking that Science knows no political boundaries. If you look through the names in these two chapters, you will see that our knowledge of the Solar System, as the Sun's family is called, has been built up step by step by representatives of Poland, Denmark, Germany, Italy, England, France and America.

Table of the Solar System

Name and symbol	Distance from Sun in millions of miles	Period of revolution around Sun	Diameter in miles	Period of rotation on axis	Number of satellites
Sun ☉	—	—	864,000	25⅓ days	—
Mercury ☿	36	88 days	3,000	88 days	0
Venus ♀	67	225 ,,	7,600	225 days*	0
Earth ⊕	93	1 year	7,927†	23 h. 56 m.	1
Mars ♂	142	1 yr. 322 d.	4,200	24 h. 37 m.	2
Jupiter ♃	483	12 years	88,700†	About 10 h.	12
Saturn ♄	886	29 ,,	75,100†	10 h. 14 m.	9
Uranus ♅	1783	84 ,,	30,900	About 10¾ h.	5
Neptune ♆	2794	165 ,,	33,000	15 h. 40 m.	2
Pluto	3666	248 ,,	About 3,650 miles	Unknown	—
Moon ☽	From Earth 239,000 miles	Around Earth 27⅓ days‡	2,160	27⅓ days	—

* Uncertain. † At the Equator.
‡ With respect to the stars. New Moon to new Moon is 29½ days.

Exercise. To make scale models of the planets. A football ('soccer') can be used to represent the Sun. Its diameter is 9 in., and this represents 866,000 miles. Hence your scale is $\dfrac{866,000}{9} = 96,000$ miles to 1 in. From the table above calculate the diameters and distances of the planets on this scale. Then make spheres of these sizes in wax or plasticene and mount them on long pins stuck into corks. The ring of Saturn (see p. 73) may be taken as 95,000 miles internal and 170,000 miles external diameter, and a scale ring cut out from paper can be attached to a pin stuck into the planet. The Earth should have the Moon at scale distance from it; this may be attached to a wire stuck into the planet.

CHAPTER XII

TELESCOPES

THE invention of the telescope is popularly attributed to Galileo, but, as indicated in chapter **x**, although he did make a telescope his was not the first. The inventor appears to have been a spectacle maker named Hans Lippershey, whose first instrument was produced shortly before 1610.

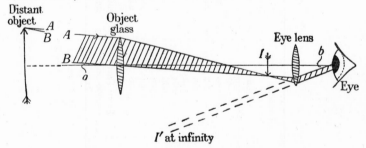

FIG. 55. The principle of the telescope

A simple astronomical telescope consists of two convex lenses, an object glass of large diameter and long focal length, and an eye lens smaller in both respects. Light from a distant object enters as a parallel beam and forms a real image at I (Fig. 55), in the focal plane of the object glass. The eye lens serves as a magnifying glass to view this image; in the diagram it is shown with its focus at *I* and the light emerging as a parallel beam. The telescope is said to be in normal adjustment, and when this light enters the eye a virtual image is seen apparently at a distance. A casual observer would focus the telescope so that *I* was just within the focal length of the eye lens, rather like Fig. 13, thus bringing the image to reading distance, but this might result in eyestrain if a constant watch had to be kept for a long time. The eye estimates size by the angle subtended by the object, which is the angle made at the eye by rays from the ends of the object. In Fig. 55 angle *a* is the angle subtended by half the object before the light has passed through the telescope, and angle *b* is that subtended by half the image. The magnification is equal to *b/a*, and can be proved to be the focal length of the object glass divided by that of the eye lens. Note that the length of the telescope is the sum of the focal lengths.

Eyepieces. If you examine a telescope you will probably find it not quite as simple as the foregoing. In place of an eye lens there is an eyepiece consisting of two lenses about an inch apart in a little brass tube, and this gives a better image than a single lens could. You will have noticed that this instrument gives inverted images. This is no disadvantage for astronomy, but it would be very awkward for a sailor or a spotter at a rifle competition. Telescopes intended for common use contain an extra pair of lenses, rather like an eyepiece in appearance, which have the effect of putting the image the correct way up. This arrangement, containing four lenses in all, is called an erecting eyepiece.

Galilean Telescope. The telescope made by Galileo worked on a different principle and is shown in Fig. 56. A concave lens is used instead of the usual eyepiece, the image is erect, and the length is the difference of the focal lengths instead of their sum. Instruments of this kind are still made for certain special purposes.

The Equatorial Telescope. Astronomical telescopes are heavy and must be firmly mounted in such a way that they can be pointed to any part of the sky. The simplest form of mounting that will satisfy these conditions must have two axes of rotation, a horizontal axis to adjust the altitude, and a vertical one to turn the telescope to the right direction. This arrangement is quite satisfactory for watching shipping or for very occasional star gazing, but there is a great disadvantage if serious work is intended. The stars move over the sky in curved paths (p. 15) and in order

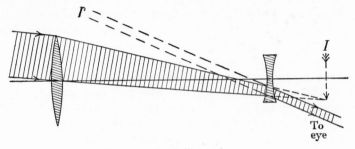

FIG. 56.   The Galilean telescope

to watch one for a considerable time you would have to alter both the direction and the altitude continually. The stars appear to move in circles around the Pole Star; if one axis of the telescope points towards the Pole Star the telescope can also be made to rotate around it. Fig. 57 shows such an instrument, called an *equatorial*, and the polar axis, pointing to the Pole Star, can be clearly seen. There is another axis at right angles to this one. If the telescope be set parallel with the polar axis it would view the Pole Star itself; if at right angles to it the telescope would follow a star on the equator of the sky when the axis was rotated at a suitable speed. Thus the equatorial mounting allows the telescope to be pointed to any part of the sky, and will then follow the object by rotation of the polar axis alone. In observatory instruments the rotation is done by machinery controlled in a manner similar to that of a gramophone.

Astronomical photography is done with an equatorial, the plate being put in the focal plane of the object glass so that the telescope becomes a camera, or else a special camera is fitted in place of the eyepiece, as shown in Fig. 57. The stars are faint, and time

Fig. 57. A 5-inch equatorial telescope equipped for photography

Fig. 58. A 4-inch transit circle

exposures, sometimes very long ones, are necessary. The equatorial mounting renders these long exposures possible by keeping the stars concerned in the same places on the plate for as long as may be desired. The mounting with horizontal and vertical motions, called an *altazimuth*, would be unsuitable even if it could be correctly driven. Fig. 59 gives two views of the Pleiades seen with an altazimuth, A when they are rising and B when they are setting. As they move over the sky they appear to rotate; the equatorial rotates the telescope in the same way and so compensates for this

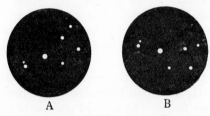

A   B

Fig. 59. The Pleiades rising and setting

movement. These long exposures are very important, for starlight that is much too faint to affect the eye will affect the photographic plate when it is allowed to fall upon the same spot for hours, and therefore the photograph can show us things that the eye, even with a telescope, could never see.

The Transit Instrument. You learnt in chapter VI that time was measured by observing when the Sun or stars cross the meridian. The telescope used in the observatory for this purpose is called a transit instrument or a meridian circle (Fig. 58). It can rotate about a horizontal axis only, so that the altitude can be altered but the direction is fixed along the north and south line. There are cross wires in the eyepiece, illuminated at night, and by watching the star and wires the exact instant of transit can be determined. The transit circle is also used to fix positions on the star sphere. The sidereal time (p. 29) of the transit of a star is called its *right ascension* and the angle of it north or south of the celestial equator (p. 42) is its *declination*. These two measurements, both of which can be obtained with the same instrument, fix the position of the star and enable it to be correctly inserted on a star map. If the R.A. of a star is known, its transit can be used to test the accuracy of the sidereal clock, and hence of clocks in general.

The Observatory. To house a large telescope an observatory is required. This usually consists of a round building with a dome. The pillar of the equatorial, containing the driving mechanism, stands on a firm foundation in the middle, and in the dome there is a shutter through which to look. The whole dome rests on wheels and can be rotated so that any part of the sky can be observed. A transit instrument does not require a dome, but only a slot in the roof and walls along the meridian. Greenwich Observatory is illustrated in Fig. 60.

The Royal Observatory, Greenwich, was founded in 1675 by King Charles II, and John Flamsteed was appointed to be the first Astronomer Royal. The original purpose of the observatory was to assist navigation, and this work is still done. It includes the determination of time (p. 27), the testing of ships' chronometers (p. 35), and the provision of material for the *Nautical Almanack* (p. 36). A great deal of research work of all kinds is carried on here in addition to the routine work for which it was founded. The largest telescope at Greenwich is 36 in. in diameter, but is a reflecting instrument, having a mirror instead of an object lens. Lens telescopes as described in this chapter are called *refractors*, and the largest of these at the observatory is 28 in. in diameter. The present Astronomer Royal is Sir Harold Spencer Jones, F.R.S.

The original observatory, built by Sir Christopher Wren and now called Flamsteed House, can be seen through the trees on the extreme right of Fig. 60, with the time ball (p. 36) prominently mounted on one of its turrets. The dome in the centre of the picture houses the astrographic equatorial, an instrument used for making photographic star maps, and a smaller equatorial is under the dome on the left. Between these two and marking the meridian of Greenwich (Long. 0°) is the transit room, of which the white painted vertical shutters facing north are clearly visible in the

A

B

Fig. 60. Greenwich Observatory: A, general view; B, the dome of the 28-inch telescope

By courtesy of the Yerkes Observatory

FIG. 61. The 40-inch Yerkes refractor

photograph. Beside the gate there is a 24-hour clock showing exact Greenwich Mean Time. The dome of the 28 in. refractor is just off the picture to the left and is illustrated separately; its peculiar shape is due to the fact that the tower was built for a much smaller telescope. The big reflector is in a newer building in another part of the grounds. The growth of London, bringing with it smoke, artificial lights and electric trains, has been an increasing handicap, so our national observatory is being slowly transferred to Herstmonceux in Sussex.

The largest refracting telescope in the world is at the Yerkes Observatory, University of Chicago, U.S.A., and was constructed in 1897 (Fig. 61). Its 40 in. object glass is likely to remain the largest, as there are various difficulties in the making of very large lenses and the modern giant telescopes are all reflectors, which will now be described.

Reflecting Telescopes. In order to overcome defects in the telescopes of his time, for the achromatic lens had not then been invented,[1] a reflector was designed and made by Sir Isaac Newton, his little instrument now being preserved in the rooms of the Royal Society. Sir William Herschel (p. 58) also made reflecting telescopes, though of a slightly different kind. Reflecting telescopes use a concave mirror in place of an object glass, and the real image produced is examined with an eyepiece similar to that of a refractor. There are several different arrangements of the mirror and eyepiece, but only the simplest, that of Newton, will be described here (Fig. 62). Parallel light from a star is reflected from the mirror M and an image is formed at F, in the focal plane. This image could be received, if required, on a photographic plate. For visual work a plane mirror intercepts the rays at m and the image is then formed at I, where it is viewed with an eyepiece E in the usual manner. A small portion of the large mirror M is obstructed by the plane mirror m, but this only dims the image a little and does not otherwise spoil it. Other types of reflector have a convex or concave mirror at m, reflecting the light down the tube again and through a hole in the principal

[1] Dollond, 1758.

mirror *M*. The mirrors used to be of polished metal, an alloy called speculum, and then this gave way to glass ones silvered on the upper surface. Many modern mirrors are coated with aluminium, and in the case of the 74 in. for the new telescope of the Radcliffe Observatory in South Africa it was intended to send it to America for aluminising on its way out from its maker in England, but the war upset the plan and it was silvered instead. The Radcliffe telescope is illustrated in Fig. 63; note the equatorial mounting, the balance weight[1] on the right, and the finder telescope in front of the observer. The largest instrument in the world was for many years the 100 in. diameter reflector of the Mount Wilson Observatory, California, U.S.A. (Fig. 64). There has recently been completed, however, after war-time delays, a 200 in. telescope on Mount Palomar, also in California. Fig. 65 illustrates a model of this giant and gives a general idea of its appearance. The model was made by the demonstrator shown in the photograph, Mr Samuel

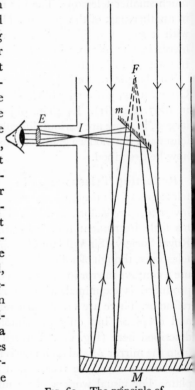

FIG. 62.   The principle of the reflecting telescope

Orkin, and is one twenty-fourth full size. The great instrument itself, the result of the inspiration and enthusiasm of the late G. E. Hale, is a mechanical marvel as well as an optical masterpiece. The huge mirror and its tube weigh nearly 150 tons, and then there is the massive U-shaped mounting; yet

[1] Or 'counterpoise'.

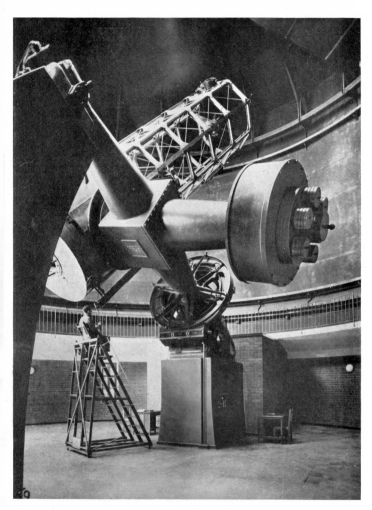

Fig. 63. The 74-inch Radcliffe reflector

By courtesy of the Royal Astronomical Society

FIG. 64. The 100-inch reflector at Mount Wilson

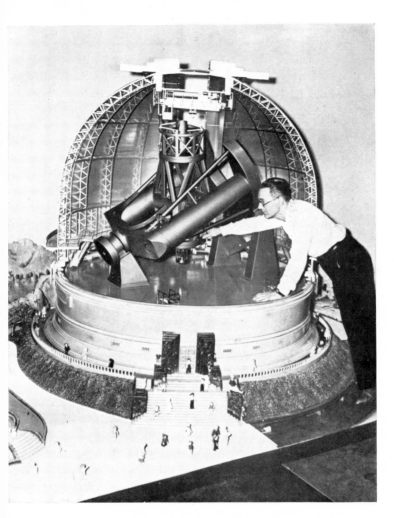

Fig. 65. Model of the new 200-inch telescope

these moving parts must work so smoothly that a star can be followed. 'The job is about equivalent to that of holding a gun in perfect aim on an object 2 inches in diameter 20 miles distant. Having in mind the size and weight of the gun, remember that its target is moving at the rate of 3 feet a second and that it has to be kept dead on its object for hours at a time during the taking of a photograph. Automatic gear has been devised that will set the telescope on any desired star *and keep it there.*'[1]

Possibly you may wonder why there are no giant telescopes in the British Isles. The climate is against us and it was always assumed that a very costly telescope would be a waste of money. Opinions have now changed, and in due course a 100 in. reflector will be erected at the new observatory at Herstmonceux. Our lack of large telescopes also explains why hitherto we have been without an aluminising plant suitable for the 74 in. mirror mentioned above.

OBSERVATION

(22) To make a simple telescope.[2]

Object glass: a convex lens about 2–2½ in. in diameter and 12–24 in. in focal length (cost, about 5s. 9d.).

Eye lens: convex, about 1 in. in diameter and 1–2 in. in focal length (cost, about 4s. 3d.).

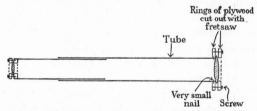

FIG. 66. To make a simple telescope

Tube: a cardboard tube about 2 in. in diameter and 2 or 3 in. shorter than the sum of the focal lengths. Such a tube can usually be obtained from an art shop, as they are used for sending

[1] F. Hope-Jones, in the *Journal of the B.A.A.* vol. L.
[2] The construction of a more elaborate telescope is given in *The Science Masters' Book*, Series II, vol. I.

prints and posters by post. You must also obtain, or make from thick brown paper, a shorter tube that will slide smoothly inside or outside the other. You should be able to devise means of fixing the lenses to the ends of tubes; Fig. 66 is given just as a suggestion. This telescope will not give perfect definition, and bright objects may appear slightly coloured, but it will be good enough to show the satellites of Jupiter, possibly the phases of Venus, and a number of the objects described in later chapters. Hold the eye lens about an inch away from the eye.

CHAPTER XIII

MORE ABOUT THE SOLAR SYSTEM

AFTER a digression upon the construction of instruments, let us return to the Solar System and see what extra information the instruments can give. We shall now let the telescope take us on a journey in space, visiting first the Moon, and then each of the planets, beginning with Mercury and working outwards from the Sun. The Sun is important enough to have a chapter to itself later on.

The Moon. The most noticeable feature of the Moon, visible to the naked eye, is the large dark patches. The ancients thought them to be seas and, although the telescope shows them to be plains, the ancient fanciful names are still in use. The smallest of instruments show the Moon's surface to be marked with a large number of ring mountains, called craters, because of their resemblance to volcanoes. How they were formed is a question not fully answered; they are probably volcanic in origin, but on account of their great size they cannot be simple volcanoes as we know them on the Earth. Although a slight eruption on the Moon has been suspected within the last century, it is fairly safe to say that her surface has been undisturbed for a very long time. The craters are not all alike, and at least three types can be seen in Figs. 67 and 68. There are ringed plains, like Plato, giant craters with a central mountain, like Copernicus, and little craters which are hollows without much mountain around them. As an indication of the size of these ring mountains Copernicus may be mentioned; it is 56 miles in diameter and the ring is some 12,000 feet above the central floor. Another type of object that can be seen

FIG. 69. Key map of the Moon

n the photographs is ranges of mountains, of which the Apennines, ing to 20,000 feet, are the most striking, and finally at full Moon markable series of rays can be seen around some of the craters, bly Tycho and Copernicus. The Moon has very little, if any, phere, and in the absence of wind, rain and rivers her moun-ave retained their rugged broken nature to a much greater extent than ours. There is reason to believe that the surface consists mainly of volcanic ash, and we know that during the day, equal in length to our fortnight, it must get very hot and during the night extremely cold. Under these conditions the existence of vegetation and animal life is almost impossible. When an occultation is being observed it is found that the stars disappear and reappear quite suddenly, which they would not do if our satellite were surrounded by a reasonably dense atmosphere comparable with our own.

Mercury. The motion of this planet is similar to that of Venus, described on page 52, and like this next door neighbour it shows phases. It is, however, nearer to the Sun than Venus, and in consequence is very difficult to see. You must look in the right place at the right time, and have the luck of a fine evening and clear sky; this set of conditions is hard to satisfy and hence to see this little planet is quite an achievement. Very little is known about it, but there is good evidence to show that it has no atmosphere, it always keeps the same face to the Sun, and has in the light and dark hemispheres extremes of hot and cold horrible to contemplate.

Venus. Probably the planet most like our own, though, being nearer the Sun, it will be considerably warmer. Unfortunately we can see nothing of its surface, for it is hidden beneath a dense atmosphere. This atmosphere is rather different from ours, for as far as we can tell it does not include oxygen, but does contain a large proportion of carbon dioxide. The phases of Venus, the most interesting feature in small telescopes, have been mentioned in an earlier chapter, but Fig. 73 is included to remind the reader of the explanation of this phenomenon. The planet is at its brightest when a crescent.

Mars. A very well-known planet, for so much has been written about it and so many speculations made as to the possibility of it being inhabited. Fig. 70 is a drawing made by a very famous amateur astronomer, the late T. E. R. Phillips, at a time when the planet was particularly well placed for observation. It is smaller and cooler than the Earth, and has a somewhat thin atmosphere which, like that of Venus, is without oxygen in

FIG. 70. Mars in September 1924

FIG. 71. Photograph of Jupiter.

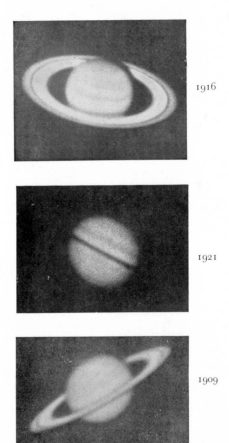

1916

1921

1909

Lowell   Observatory

FIG. 72. Photographs of Saturn

appreciable quantity. We can see its surface and watch the seasonal changes that take place in the Martian year, equal to about two of our years. At the poles there are white caps suggestive of polar snows, and these diminish in the summer and increase in the winter. The rest of the surface is of a brownish colour, with some variations in tint showing seasonal changes, and the planet shines with a reddish light. It is called the 'ruddy planet', and no doubt its colour has some bearing on its name, Mars, the god of war.

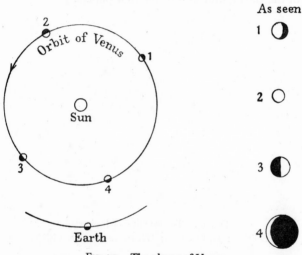

As seen

FIG. 73. The phases of Venus

The so-called 'canals' of Mars are a number of straight dark markings that have been seen by some astronomers and whose existence is disputed by others. If they do really exist, they must be regarded, on account of their size, as natural features, such as areas of vegetation. Mars has two very small satellites, Phoebus and Deimos, revolving around it at comparatively short distances and in the very short times of $7\frac{1}{2}$ and 30 hours respectively.

Minor Planets. If you examine the table of distances from the Sun (p. 59) you will see that the first four planets are fairly evenly spaced, but then comes quite a large gap. This does not mean a great empty space, for between the orbits of Mars and Jupiter lie those of well over a thousand minor planets or asteroids. These bodies are very small, the largest of them, Ceres, being 485 miles in diameter, and their average size only about 40 miles. They are not visible to the naked eye, with the possible exception of Vesta, which is the third largest. One of the asteroids, Eros, became quite well known at the beginning of 1930, for owing to its orbit being very elliptical it came to within about 15,000,000 miles of the Earth, and so into the newspapers.

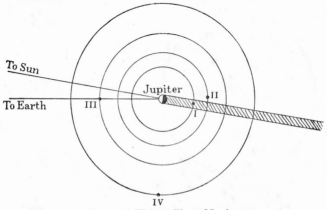

FIG. 74.　The satellites of Jupiter
I, Eclipsed; II, Occulted; III, In transit; IV, Visible.

Jupiter. This is the largest planet, being 88,700 miles in diameter at the equator and 82,800 at the poles. In chapter II it was mentioned that the Earth is slightly flattened at the poles; Jupiter is very much so, as shown by the figures above, and in fact the slightly oval shape of its disc can be seen with quite small telescopes. The flattening is an indication of very rapid rotation, and its period of just under 10 hours is the shortest of all the planets. We cannot see its surface, but only its atmosphere, which is believed to be several thousand miles thick, making the solid planet considerably smaller than the measurements above. The atmosphere shows clouds in clearly defined but varying belts (Fig. 71), though on account of the very low temperature they

cannot be water clouds, like ours, but must be of some light con-
densible gases such as ammonia. Mention has already been made
of the four large satellites discovered by Galileo. They can be seen
with very small telescopes as tiny points of light forming, with
Jupiter, very nearly a straight line. They are interesting to watch,
as their arrangement changes from night to night, and some movement can be detected in the course of a single evening. They are
frequently not all visible at the same time, as they may be in front
of the planet, behind it, or in its shadow, as illustrated in Fig. 74.
By observing the eclipses of these satellites the velocity of light has
been determined; this will be explained later on. Jupiter has also
eight smaller satellites, one being discovered as recently as 1951.

Saturn. The next planet is also the next in size and in itself
is very similar to Jupiter. The chief claim to fame in this case is
the very remarkable ring, the only known example (Fig. 72). This
ring, or series of rings, usually considered to be three in number,
is not rigid, but consists of a whole host of tiny particles moving
around the planet like satellites, and is probably the remains of
an ordinary satellite which has broken up. The outside diameter
of the ring system is about 170,000 miles, but it is very thin indeed,
so that when it is exactly edge on to the Earth it is almost invisible.
The rings do not lie quite in the plane of the ecliptic, so we sometimes see the north side of them and sometimes the south. In
Sept. 1950 the rings were edge on, and we now see the north side,
the rings next being open to their greatest angle early in 1958. In
addition to the rings Saturn has nine satellites.

Uranus and Neptune. These planets are so far away that
the telescope tells us little of them except that they are of the same
general character as Jupiter. Uranus has five satellites and
Neptune two; most of them have orbits making big angles with
the plane of the ecliptic and they appear to move around their
planets in the wrong direction, i.e. in a direction opposite to that
in which most members of the solar system move.

Pluto. Of this little planet we know practically nothing,
though it is believed to be small. For the present we may think
of it as being something like Mars but very, very cold.

Comets. We do not often see a really spectacular comet, but one can quite imagine that to the ancients it was an object of fear and awe. In Fig. 76 you see a bright head and a long shining tail, but the comet that aroused some interest in 1943 was only an ill-defined star just visible to the naked eye, and this was brighter than most. Records of bright comets go back for very many centuries, and on examining those records Halley (1656–1742) decided that comets were periodic, i.e. that they returned to visibility at regular intervals. He studied the

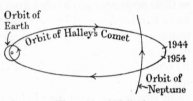

FIG. 75.   The orbit of Halley's comet

motion of the comet of 1682 and predicted that it would be seen again in 1758. Quite near to his predicted date it returned, thus confirming its period of 76 years. It was last seen in 1910, and by working backwards it has been found that this was the comet that worried King Harold at the time of the Norman Conquest and is shown in the Bayeux tapestry. Comets belong to the solar system, but their orbits are very long ellipses as shown in Fig. 75, and the part near the Sun, where the planet is under observation, is hardly distinguishable from a parabola, which is the path taken by a cricket ball when a high catch is hit. A comet is quite light compared with a planet, for the head is no more than a cluster of small fragments. In the remote part of its orbit it moves very slowly and is invisible from the Earth, but as it gets nearer the Sun it becomes brighter and faster and for a short period is seen among the stars. The tail, if any, only appears when the comet is near the Sun, and always points away from that body. Halley's comet passes out about as far away as Pluto; some do not go so far, while others travel well beyond the limits of the rest of the Sun's family. Some have been known to change their orbits. The gravitation between the Sun and comet determines the original orbit, but it may happen that the comet passes so near to a planet that for a short time it is pulled out of its path by the force of gravity of that body. As the distance from the planet increases again the Sun regains control, but the orbit will not be the same as it was before.

Lowell Observatory

Fɪɢ. 76. Halley's comet in 1910

Meteors.   Shooting stars are familiar to everyone, that bright streak across the sky leaving, sometimes, a trail of sparks. They are small particles of matter, sometimes stony and sometimes metallic, travelling through space, and when near the Earth they fall towards it. Their great speed through the atmosphere causes friction and a big compression of the air in front, with the result that they get white hot and burn away. There is great variation in their speed and height, but as a rough average the speed could be quoted as 20 miles per second and height about 50 miles. Sometimes the piece of rock is an exceptionally large one; the trail is then much more brilliant and the meteor usually terminates its existence in a loud explosion. This phenomenon is called a *fireball*. Very occasionally they fall to Earth as *meteorites*, or meteoric stones, and it is fortunate that large ones are rare, as much damage might be done if one fell in a populous place. A large meteorite which fell in Siberia in 1908 felled and scorched the trees for miles around. There are certain dates on which meteors are particularly numerous and appear to come from the same part of the sky, called the radiant. There is an important[1] meteor shower in November, called the Leonids because the radiant is in the constellation of Lion, and at the beginning of August many shooting stars may be seen coming from the direction of Perseus. These showers are due to shoals of meteors travelling through space in cometary orbits, and some, indeed, are known to be the remains of former comets.

The Velocity of Light.   Let us now return to Jupiter's satellites, the eclipses of which have already been mentioned. By observing their motion carefully the intervals between the eclipses can be calculated and hence future eclipses predicted. Suppose an eclipse is observed when the Earth and Jupiter are at E_1 and J_1 respectively (Fig. 77). Working from that the exact time can be predicted for an eclipse about six months later, when the planets are at E_2 and J_2. When the time comes, however, the eclipse will occur about 1000 seconds late. We see an astronomical event only when light emitted at the moment of occurrence has reached our eyes, and in the second of these eclipses the light has farther to travel than in the first, for $E_2 J_2$ is longer than $E_1 J_1$ by the diameter

[1] By past reputation; a brilliant display has not occurred for many years.

of the Earth's orbit. Thus in 1000 seconds light travels 2 × 93,000,000 miles, and its velocity is therefore 186,000 miles per second. This was the first successful method of finding the velocity of light, and

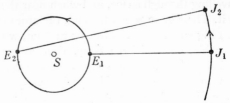

FIG. 77. The measurement of the velocity of light

was used by Roemer, a Danish astronomer, in 1675. Since his time terrestrial methods have been developed by which the velocity can be determined over distances measurable in yards instead of millions of miles.

OBSERVATIONS

(23) Examine the Moon with a small telescope, and try to identify the objects named in Fig. 69.

(24) From *Whitaker's Almanack* find out when Mercury is likely to be visible and look for it. As an evening star it will be low in the western twilight soon after sunset.

(25)* If a telescope is available, try to see the phases of Venus. You should observe it before the sky is really dark, in fact as early in the evening as you can find the planet. If your instrument is a very small one, you will be most likely to be successful when Venus is a crescent, which is just after elongation of the evening star or before it in the case of the morning star.

(26)* With a small telescope look for the satellites of Jupiter. Make a drawing of their arrangement every available night for several weeks.

(27)* If you have a telescope of not less than 2 in. diameter, look for the rings of Saturn.

(28) Watch for meteors regularly in the constellation of Perseus between July 25 and August 15, August 10 and 11 being the most important nights. The best time for the Leonid meteors is after 1 a.m. on November 14 and 15.

* Find suitable dates as for Observation (24).

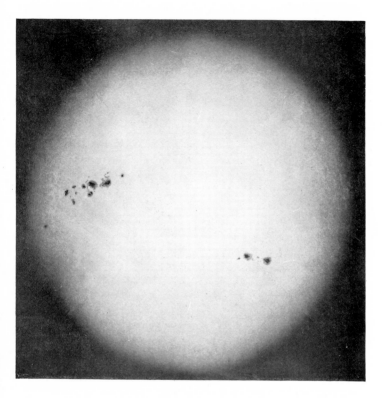

Fig. 78. The Sun, showing sunspots

FIG. 79. The solar eclipse of 1919, showing the famous 'ant-eater' prominence

CHAPTER XIV

THE SUN

THE SUN is a glowing sphere 864,000 miles in diameter. It is the parent of the Solar System, the body from which we believe the planets came, and which now sustains them with its light and heat. We see only its *surface*, and the nature of this we now know quite well; the nature of the Sun's interior is a problem not yet fully solved. When the Sun is photographed or examined with a telescope, taking due precautions to protect the eyes, the surface appears mottled, almost like rice pudding, and can be seen to be in a very disturbed state.

The ancients regarded the Sun as something perfect, and when Galileo announced that he had seen spots on its surface, he was disbelieved by most learned people of his time. Actually *sunspots* had been recorded by the Chinese many centuries earlier, for occasionally they are large enough to be seen with the naked eye when the Sun is dimmed near sunset. With a very low-powered instrument they are just black spots; with a higher power, one of these shows a black spot, the umbra, surrounded by a greyish penumbra, and thus looks like a funnel-shaped hole in the Sun's surface. They come and go, some in a few hours, others in a few weeks. Sometimes the Sun has many spots and sometimes none. Astronomers have observed and measured the spots for many years, and as a result they have established an eleven-year cycle. From year to year the spots gradually become more numerous and then diminish again, the whole process taking 11 years. The last maximum, a particularly high one, occurred over some months in 1947–8, and thus the last few years have been a period of decreasing activity. Individual spots vary greatly in size, but spots large enough to put the Earth in without touching the sides are by no means unusual. Sunspots do not occur all over the disc, but in zones corresponding roughly with the temperate zones of the Earth. By watching daily, groups of spots which persist for a time can be seen crossing the disc, thus showing that the Sun rotates on its axis. The time of rotation is a little over 25 days; this is an average value, for the equatorial regions rotate more rapidly than the higher latitudes.

A total eclipse reveals another important feature, Fig. 79, for when the disc of the Sun is covered by that of the Moon we can see the light of the outer layer of the former. Under these conditions *prominences* can be seen. They are huge masses of incandescent gas, largely of hydrogen and flame-like in appearance, though they ought not to be called flames. They are in constant movement and are sometimes several hundred thousand miles long. The other delicate light that can be seen around the eclipsed Sun is called the *corona* (also visible in the frontispiece), believed to consist of very minute particles, some proportion being probably electrically charged.

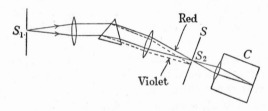

FIG. 80.　The principle of the spectroheliograph

It is possible to observe prominences without waiting for an eclipse, the ever useful spectroscope being applied. Suppose that the primary image (*I* in Fig. 55) of the Sun due to a telescope be directed into a spectroscope, the edge of the disc falling exactly along the slit. The full glare of the Sun will be spread out over the whole spectrum, but the hydrogen light from a prominence, if there happens to be one, will be concentrated into the usual line spectrum and becomes visible. If now attention be concentrated on one line, usually the red one, and the slit be widened a little, the whole prominence becomes visible. A more elaborate application of the spectroscope enables photographs of the Sun to be taken in any spectral light desired, the apparatus in this case being

FIG. 81. The Sun, photographed in hydrogen light

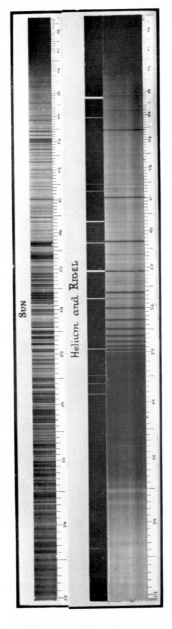

SUN

Helium and RIGEL

Fig. 82. Photographs of spectra. Ultra-violet on the left, blue on the right

called the spectroheliograph. Fig. 80 illustrates the principle only and not an actual instrument.[1] An image of the Sun falls on the slit S_1, which is therefore illuminated by one narrow strip of the image. The spectrum formed on the screen S consists of images of this strip in every light that is present, and any one of these images, say that due to the red hydrogen line, can be examined separately by putting the slit S_2 on it. Now if the image of the Sun be moved over S_1 the whole of it will be seen, in turn, by hydrogen light only and can be photographed by the camera C. In practice the image of the Sun remains stationary and the slits move, S_1 across the disc and S_2 keeping step with the hydrogen line as the spectrum moves over the screen S. Fig. 81 shows the Sun in hydrogen light. The bright markings are due to masses of hydrogen called flocculi; the dark marks are due to absorption by hydrogen above the Sun's surface, i.e. to prominences; this particular photograph does not show prominences on the edge of the disc.

Now let us turn to the ordinary spectrum of the Sun. At first sight it is just a continuous spectrum, but a good spectroscope shows it to be crossed by scores of narrow dark lines, shown in the upper photograph of Fig. 82. These are called *Fraunhofer's lines*, after the German scientist who, in 1814, gave them their meaning, though not their full explanation. With a suitable instrument it is possible to examine two spectra at the same time, and if those of sodium and the Sun be so examined it is found that the bright line of sodium occupies the same place in the spectrum as one of Fraunhofer's lines. The same is true of the hydrogen spectrum and many others. Use a ruler and compare these spectra in Fig. 15; this diagram is greatly simplified, of course. The continuous spectrum is due to light from the interior of the Sun, and the absorption lines are caused by the elements in the outer layers through which the light has passed. The vapours in the Sun's outer layers are emitting their usual lines, of course, but this light is not enough to make up for the absorption. The bright lines can be seen, however, on the occasion of a total eclipse, when the central glare is hidden but, just for an instant, the outer layer is not. The formation of

[1] The arrangement of S_1, prism and lenses shown here, is commonly used for studying pure spectra in the laboratory.

Fraunhofer's lines can be imitated in the laboratory,[1] so there is really no doubt that the foregoing explanation is correct. Thus Fraunhofer's lines show what elements the solar atmosphere contains, and the spectroscope also provides information from which the temperature can be determined, about 6000° C. The temperature of the interior must be many times hotter than this.

Another interesting point about the solar spectrum is that the gas helium, now used for filling airships, was found in the Sun before it had been found on Earth. Certain lines could not be identified with any known element, and were supposed to be due to one which occurred only in the Sun; hence the name helium (*helios*, Greek for Sun). Years afterwards a gas giving this spectrum was detected as a constituent of natural gas issuing from the ground in North America, and its lightness, second only to hydrogen, has made it a product of commercial importance. It also occurs in small quantities in certain minerals and in the atmosphere.

The Sun has several effects upon the Earth other than those of light and heat. There is, for instance, a connection between our climate and the 11-year cycle. The Earth's magnetism, that force which causes a compass to point in a northerly direction, sometimes shows remarkable variations spoken of as *magnetic storms*. These phenomena tend to occur at times when there are many or extra large spots on the Sun, and it is thought to be due to some form of electrical energy emitted by the spots. The approach of a spot maximum is regarded with some anxiety by the users of long-range short-wave wireless, as this is interfered with by magnetic storms. Associated with the magnetic storms, and frequently occurring at the same time, is the *Aurora*. This is a beautiful glow seen high up in the Earth's atmosphere in the polar regions, and possibly similar in nature to the solar corona.

Zodiacal light is a different phenomenon altogether, being best seen by observers in the tropics. It is a very pale luminosity sometimes seen in the west after sunset or in the east before sunrise, and is usually in the form of a narrow cone pointing along the ecliptic, its widest and brightest part being on the horizon. The suggested explanation of this is that the space round the Sun is occupied by

[1] *Science Masters' Book*, Series II, Part 1, Expt. 123.

a disc of very fine dust, the diameter of the disc being comparable with that of the Earth's orbit but the thickness small. It might be imagined to be something like Saturn's ring system, but continuous instead of in rings and composed of very much smaller particles of matter. Scattered light from this dust is just visible on a dark clear night as the Zodiacal cone under discussion.

OBSERVATIONS

(29) If a telescope is available, examine the Sun on every fine day for at least a month. Draw a diagram showing the positions of the spots each time, and from your diagrams try to deduce the period of rotation of the Sun. *On no account look directly at the Sun*

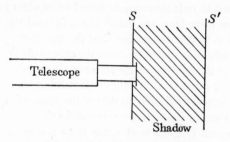

FIG. 83.   Method of observing the Sun

through the telescope; use this method of observation: Attach a cardboard screen, S (Fig. 83), to the eye end of the telescope so that there is a large shadow cast by it when the telescope is pointed towards the Sun. Hold a white screen at S', from 6 in. to a foot away from the eyepiece, so that the sunlight passing through falls upon it. By adjusting the focus of the telescope a sharp image of the Sun can be obtained on S' and the spots observed easily and safely.

(30) On clear moonless nights during the period February to April inclusive, examine the western sky for Zodiacal light as soon as darkness sets in.

CHAPTER XV

THE STARS

THE STARS look to us just like tiny spots of light. With the naked eye we notice a great variety in brightness and colour; a telescope just makes them look brighter, but on account of their very great distance it cannot make them look larger. One star alone is near enough for us to see in detail, and that is the Sun. The Sun is just an average star. Of the others some are larger and some smaller; some are hotter and some not so hot; some are whiter and some more inclined to red; some are believed to be older and some not so old. How these facts are found out is beyond the scope of this book, but it is well to remember that the spectroscope is the chief instrument of stellar research. As in the case of the Sun's spectrum, that of a star shows what elements are present in its outer layers, but it gives a great deal of other information too. Some stellar spectra are generally similar to that of the Sun, while others, such as that of Rigel in Fig. 82, are quite different.

The apparent brightness of a star is spoken of as its *magnitude*. The brightest stars, shown as large dots in Figs. 22 and 23, are of the 'first magnitude'. Stars not so bright, such as all but one of the Plough, are of the second, the next brightest the third, and so on, each magnitude being about two and a half times as bright as the next magnitude fainter. The naked eye can see stars down to the fifth or sixth magnitude, ordinary binoculars to the seventh or eighth, and photography through a giant telescope reveals stars of the twenty-first magnitude.

An instrument for measuring, or perhaps we should say comparing, the brightness of stars is called a *photometer*, but much good work is done, particularly by amateurs, by eye estimation alone. Magnitudes are sometimes determined from the images on a photographic plate instead of from the stars themselves, though owing to the variations in the colour of stars photographic magnitudes are not quite the same as visual ones.

The stars are not all of constant brightness, and *variable stars* are quite numerous. An important example is Algol, in the constellation of Perseus. Normally it is seen to be of the second magnitude, but every three days it fades in about four hours to fourth magnitude, and then in the next four hours regains its normal brightness. The actual period from minimum to minimum is 2 days 21 hours. The explanation is that Algol consists of two stars, one bright and one very faint, revolving around one another, and every three days an eclipse takes place, the dark one getting in front of the bright one and cutting off its light. There are several other types of variable stars besides the eclipsing ones. An important class, called the Cepheids, have short periods like Algol, but their variation appears to be due to some change in the star itself, and not to eclipses. Others have long periods measurable in years and months rather than in days and hours, and a further class is those which have no period at all and just vary irregularly. Finally there are the *Novae*, or new stars, where a very faint star suddenly becomes bright, or a star appears in a place where no star had been noticed before. A nova appeared some years ago in the constellation of Hercules, not very far from the star Vega in Fig. 22. On December 13, 1934, it was noticed by an amateur astronomer and was then quite faint, but by Christmas it was not far short of the first magnitude. With irregular variations it remained visible to the naked eye until the beginning of April 1935, when it quite suddenly faded. This star had been of the fourteenth magnitude prior to this mysterious outburst, and by 1941 it had dropped back again into obscurity, being reported to be of the thirteenth. Nova Puppis of 1942 was so low in the English sky that, although quite equal to its predecessor above, it did not arouse so much general interest. The cause of outbursts like this is not definitely known.

To some extent the great variety in brightness that we see on a starry night is due to variations in distance. A faint star is not necessarily a distant one, but as a rough general rule we can say that the very bright stars, like Sirius, are comparatively near to us, and the very faint ones, such as those of the Milky Way, are at a much greater distance. The *distances of the stars*, even the near ones, are almost too great to imagine. Reference was made on page 60 to a scale model of the solar system using a football for the Sun. On this scale the Earth would be a small pea one cricket pitch away, and if the model Sun were in Britain the nearest star of all would be in Canada somewhere near Winnipeg! The actual distance of the nearest star is approximately 25,000,000,000,000 miles, and as this is such a large figure it is more usual to use a unit called a 'light year', the distance light travels in a year.

Light travels 186,000 miles in a second;

or　186,000 × 60 miles in a minute;

or　186,000 × 60 × 60 miles in an hour;

or　186,000 × 60 × 60 × 24 miles in a day;

or　186,000 × 60 × 60 × 24 × 365¼ miles in a year
　　= 6 billion approximately.

Hence a light year is 6 billion miles, and the distance of the nearest star is therefore just over four of these units. A few other distances are given in the following table:

| Name | Distance | | Diameter in miles | Remarks |
	As a light journey	In millions of miles		
Sun	8 m.	93	864,000	—
Sirius	8·6　yr.	51,000,000	1,380,000	The apparently brightest star
Proxima Centauri	4·27　,,	25,000,000	61,000	The nearest known star
Betelgeuse	190·00　,,	1,120,000,000	250,000,000	A very large star
Van Maanen's Star	12·8　　,,	75,000,000	7,800	A very small star

The most distant object that the naked eye can see, called the Nebula in Andromeda (p. 88), is about 800,000 light years away, and the telescope reveals objects at much greater distances than that.

The word 'parallax' means the apparent motion of an object against its background, caused by the real motion of the observer. S represents, in Fig. 84, a star that is much nearer than the average to the Solar System, and its directions seen from the Earth are $E_1 S$ when that body is on one side of the Sun and $E_2 S$

when on the other. E_1X_1 and E_2X_2 are the directions of a star so distant that the lines can be considered to be parallel. Then it is obvious that as the Earth moves around the Sun the star S will appear to move when compared with X. This movement can be measured and from it the angle P, half of which is called the parallax of the star, can be deduced. Then, as E_1E_2 is known, the distance of the star S can be calculated; this method was first successfully used, by several astronomers, in the eighteen-thirties. These angles of parallax are very small indeed, that of the nearest star being only three-quarters of a second (1 second $= \frac{1}{3600}$ of a degree), and get smaller as distances get greater. If a star had a parallax of 1 second, it would be at a distance of 1 'parsec'; this is the unit

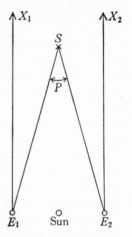

FIG. 84. The parallax of a star

used by astronomers in place of the light year. The parsec is just over three light years, or 19·2 billion miles.

The apparent brightness of a star can be utilised as a means of judging its distance, for the fainter it looks the farther away it is, but it will be necessary to know how bright the star actually is. In certain cases this 'intrinsic brightness', which for simplicity we might call its candle-power, can be calculated from spectroscopic or other information, and then if the apparent brightness be measured the distance can be calculated. It does not follow,

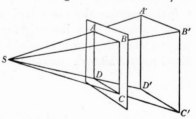

FIG. 85. The inverse square law

however, that half the brightness means double the distance, for light does not obey quite such a simple rule. Instead there is an *inverse square law*, the meaning of which you should know from chapter XI. Look at Fig. 85. S is a point source of light and

ABCD a square hole in a sheet of cardboard held so that the light falls at right angles to the middle of the hole. If a plain white screen be held behind the first a square of light will fall upon it, and as this screen is moved farther back the square will get larger but less brightly illuminated. Now suppose that $A'B'C'D'$ is just twice as far from S as is *ABCD*. Then $A'B'$ will be twice AB, and the area of the square of light will be four times that of the aperture through which the light passed. Thus the illumination of the screen will be only one-quarter of what it would be if the screen were actually in the aperture itself; it is like spreading a fixed amount of butter on four pieces of bread instead of on one. For three times the distance the illumination would be 1/9; for four times, 1/16 and so on. As for the brightness of the light S itself, our estimate depends upon the illumination of our eyes, and as the latter decreases with the square of the distance so does the apparent brightness of S. The inverse square law is 'the intensity of illumination varies inversely as the square of the distance from the source'. The Cepheid variables have proved of great value for distance determinations, for a relation has been found between their period of fluctuation and their intrinsic brightness or luminosity. Knowing this quantity and the apparent brightness we can, as indicated above, apply the inverse square law and deduce the distance.

OBSERVATION

(31) Examine the constellation of Perseus and notice how Algol compares with the neighbouring stars. Repeat the comparison when the star is at minimum; the date and time can be found from *Whitaker's Almanack*.

CHAPTER XVI

THE STELLAR UNIVERSE

THE last chapter was about stars, their colour, their brightness and their variation. The telescope, however, reveals very much more than simple stars beyond the confines of the solar system.

Now we shall deal with each of these things separately, and finally try to obtain a mental picture of the whole universe. This chapter is but the scantiest outline of a vast subject, one developed by so many workers that no attempt will be made to add to each point of interest a discoverer's name.

Double Stars. If the middle star of the tail of the Great Bear (i.e. the handle of the Plough) be carefully examined with the naked eye, a faint and close companion will be seen. Thus what is casually passed as one star is really two, Mizar, the brighter, and Alcor, the fainter. Fig. 86 (A) shows this naked-eye view, and

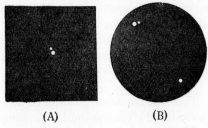

(A)   (B)

FIG. 86. Mizar and Alcor: (A) with naked eye, (B) with small telescope

in (B) you see a view in a 2 in. telescope. The pair are now wide apart and Mizar itself has become two. Mizar and Alcor form a 'naked-eye pair'; Mizar is one of a very numerous class of objects called 'double stars'. In some cases, such as Castor (Fig. 23), the components can be seen to be slowly revolving around one another, once in many years; in others high magnification shows that one or both of the components are themselves double. Close revolving pairs are called 'binary stars' and groups of three or more 'multiple stars'. The spectrum of the brighter component of Mizar is peculiar. It shows Fraunhofer's lines in the usual way, but if examined regularly they will be found to show periodic changes. Sometimes for each line there are two lines close together, and then later on they become single again. This is evidence that it is a double star too close for the telescope to separate, and it is an example of the class, also numerous, called 'spectroscopic binaries'. Summarising the facts about this star, we have Mizar and Alcor a naked-eye pair, Mizar the first discovered telescopic double star, and its brighter component the first discovered spectroscopic binary.

Star Clusters. The name describes them, a lot of stars crowded together into one compact group. It is important to realise that the stars concerned really are in a group, and not merely in the same direction. Two clusters just visible to the naked eye are marked in Fig. 23; the one near the Lion is a rather scattered group, while the Perseus one is closer packed, though coarse when compared with such an object as Fig. 87. This is a famous one in the summer constellation of Hercules, and is called a 'globular' cluster because the very close packing in the middle suggests that shape. Do not think the stars are really as near to one another as they look, for that is an effect of the distance of the cluster, which is about 36,000 light years.

Nebulae. This was the name given to certain misty patches of light, and included the clusters until their nature was known. A nebula is marked on Fig. 23 in Orion, and a photograph of this is reproduced in Fig. 88. The spectrum of this kind of nebula includes some bright emission lines, and this shows (refer to chapter 1) that the source of light is an incandescent gas. The same photograph shows the existence of a non-luminous form of matter away in space, for in several places the light ends with great suddenness and becomes a dense blackness. This is due to absorption by some other matter which is emitting no light of its own. An obscuring mass of this kind is sometimes called a dark nebula.

Like the nebula in Orion, that in Andromeda is just visible to the naked eye, and is the very distant object referred to in the last chapter. It is of quite a different class, however, and is one of the many 'spiral nebulae'. Fig. 89 at once gives the impression of a round object seen obliquely, and the spiral nature is evident. Some similar nebulae are seen from the Earth full face, looking round and obviously spiral, while others are edge on and show that these objects are thin and disc-like. Other characteristics of the spiral nebulae are (a) that they can to a considerable extent be resolved into stars, (b) their spectra resemble those of stars, and (c) they

Dominion Astrophysical Observatory, Victoria, B.C.

FIG. 87. The star cluster in Hercules

5'

Fig. 88. The nebula in Orion

Fig. 89. The nebula in Andromeda

W. H. Steavenson, Pretoria

FIG. 90. The Milky Way. This is the brightest portion; the middle of
the picture is in the constellation of Sagittarius

are very much more distant than any of the other objects considered, most of them being millions of light years away.

The Milky Way. This beautiful silvery band of light, also called the 'Galaxy', goes right around the celestial sphere, though, of course, we only see about half of it at a time. It consists of stars, very numerous and mostly very distant, and we naturally wonder why so many should be gathered into this ring. The spiral nebulae give us a clue, for if we were inside one of them we should see many more stars when looking towards the edge, direction A in Fig. 91, than towards the sides, direction B, because of the much greater depth to look through. It appears, then, that the stars occupy a disc-like volume in space, the Milky Way being the plane of the disc, and thus the number and average faintness of

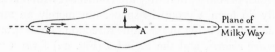

FIG. 91. The shape of the galaxy

the stars in that plane is due to the fact that it is the direction of greatest depth. There is other experimental evidence to confirm this view, and the explanation can be accepted with confidence. The size of this 'galactic system' is still in doubt, but it is probably somewhere near the truth to say that its diameter is about 150,000 light years and that it contains about 100,000 million stars. The Sun is situated quite near to its central plane, but not at the centre. S in Fig. 91 is about the place, and the arrow there shown is the direction in which we believe we are looking when we see the part of the Milky Way illustrated in Fig. 90. The galactic system includes the star clusters, the irregular gaseous nebulae, and the dark obscuring matter, but not the spiral nebulae. These last are other galaxies far outside our own. They are usually known as the 'extra-galactic nebulae'.

Stellar Motions. As a distinction from the planets the stars were once called 'the fixed stars'. This is not strictly true, though to a naked-eye observer the constellations do not change in a lifetime. By comparing star maps several hundred years old with modern ones some changes can be seen, for the stars are moving. Accurate measures of these motions can also be obtained by comparing astronomical photographs taken over a period of years, both these methods, of course, revealing motion across the line of sight. Motion in the line of sight, i.e. directly towards us or away from us, can be detected and measured by means of spectroscopic observations. These apparent motions among the stars have been studied in great detail; it will be sufficient to mention here just four points about them. The Sun itself is moving, taking the Solar System with it, at a velocity of about 12 miles a second, and in the direction of the constellation of Hercules. Allowance must be made for this motion when considering that of other stars. The next point is that certain groups of stars have the same motion. This suggests, of course, that they are connected with one another and really do form a group in space, not merely an apparent group due to their being in the same direction. These groups are called *moving clusters*, and the Pleiades (Fig. 22) form a notable example. The third point is that the Galaxy is in a state of rotation, the part near the Sun taking about 250 million years to go around once. The other galaxies show a similar motion, the period of the Andromeda nebula being about 20 million years. Finally, most of the spiral nebulae, these other galaxies, appear to be moving away from our own at tremendous speeds of hundreds, and even thousands, of miles per second. The farther off they are the faster they run away, and it is this motion that is being spoken of when you hear mention of 'the expanding universe'.

The Completed Picture. The Earth is just one of the smaller members of a group of planets revolving around a central star, the Sun. The Sun and many of its nearest neighbours are moving through space at a speed of about 12 miles a second. Although we use the word 'neighbour' the Sun's nearest fellow is 25 billion miles away. Our group, with clusters, nebulae and isolated stars, forms a part of a huge revolving disc, so large that an almost inconceivably large number of stars still remain at almost inconceivably large distances apart. Sir James Jeans gives an impression of the emptiness of space when he says:[1] 'leave only three wasps

[1] *Stars in their Courses.*

alive in the whole of Europe and the air of Europe will still be more crowded with wasps than space is with stars'. This galaxy to which we belong is not alone; it is one of some several million others, spaced from one another at distances of a million or more light years. The whole universe is truly large, yet it appears to be getting larger, for its islands are moving apart with speeds of thousands of miles a second.

The Birth of the Stars. We now leave *the facts* of astronomy and move on to *pure conjecture*. There have been many attempts to explain how the universe we see developed and to write its history, and although they appear to explain some of the facts there is at the present time no complete theory that does not break down somewhere. What follows is just an outline of some of the suggestions that have been made.

In the beginning the whole of space was filled with a very very thin nebula uniformly scattered; possibly the dark nebulae are remains of this material. Somehow its uniformity became disturbed, and wherever there was a slight accumulation of mass the force of gravitation drew other matter towards it, so that in time the uniform nebula had gathered into patches, one, our own, being larger than the rest. This process is called condensation. We shall now consider one of these patches, which would probably assume a spherical shape. Rotation began, and as condensation made it smaller and more compact the rate of rotation increased and it began to flatten. When it had become the shape of a convex lens it did not go on flattening, but pieces came off from the rim and it became a somewhat broken spiral. The pieces then behaved in a way similar to that of the original nebula, but the resulting separate masses were much smaller and a cloud of stars was produced. If this star cloud were in a congested part of the spiral nebula, it would gradually scatter; if fairly well isolated, it would take up a spherical, i.e. globular, shape. Thus the stellar universe has been accounted for, though the evidence for its past is almost entirely mathematical. Observation supports it, however, for the extra-galactic nebulae are far from being all spirals, and the sky as a whole can provide for examination examples of every stage of the evolution described.

Some stars continued their whole existence as single bodies, others became binaries. The nature of the interior of a star is a matter of considerable doubt, but assuming that it has some of the properties of a liquid the formation of a binary can be shown mathematically to be the natural consequence of rapid rotation. It flattens at first, just as we can see, for instance, the flattening of Jupiter, but, unlike a gas, it does not go on to the lens shape. Instead it becomes elongated, then it becomes narrow in the middle, breaks into two parts, and finally we have a revolving double star.

The theory of the last paragraph is less certain than that before it, for we cannot see it in its various stages, and the next suggestion, the story of an unusual accident, is even more of a conjecture. In the course of its journey through almost empty space the Sun passed quite near to another star, a large one, and owing to gravitation a tremendous tide was raised on the Sun's surface. The matter was drawn towards the visiting star so strongly that a portion, in the shape of a cigar, was pulled right out. By then the stars had separated and the Sun was left with this detached matter slowly revolving around it. It soon broke up into separate pieces, large in the middle and small at the ends, and in time these pieces cooled and contracted, and became the planets. At sometime during their early life the Sun raised tides on the planets, particularly the larger ones, and in a manner similar to their parents the satellites were born. This tidal theory is not generally accepted, but it does seem likely that the approach of another star was an essential factor.

This completes a story of the birth of the stars and solar system, and the three processes described are illustrated diagrammatically in Fig. 92. It is a very long story, and although dates cannot be given with any degree of certainty some suggestions can be mentioned. There seems to be fair agreement that the age of the Earth is about 2000 million years. The total life of the Sun has been estimated to be at least 50,000 million years, but much of this is still to come as the age of the universe, based on its expansion, appears to be only about 30,000 million years. Great as these numbers are a much longer time scale was formerly demanded.[1]

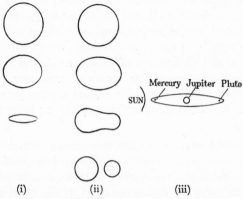

FIG. 92. Illustrating the birth of the stars: (i) Rotation of gases forming spiral nebulae. (ii) Rotation of liquids forming double stars. (iii) Planets produced by tidal action

Life in other Worlds. The possibility of life in other worlds is something about which many must have speculated.[2] It was referred to in chapter XIII, and from the descriptions of the planets there given it is apparent that of the solar system only Venus, Mars and the Moon have surfaces and temperatures which could permit life. The Moon can safely be eliminated owing to its lack of atmosphere. Venus is a hot planet with a dense atmosphere, and Mars a cold planet with a sparse one, both these atmospheres being deficient in oxygen. Thus life, if any, on these two planets must be very different from ours, and probability rather favours none at all. There may be planets revolving around other stars, but on account of distance we cannot find out with certainty, though very recently two stars have revealed evidence of the existence of a possible planetary attendant. We can, however, consider the likelihood of this being the case. We have seen that the formation of planets is possibly due to the close approach of

[1] The shorter period is given in the 1944 edition of Jeans' *Universe Around Us*; in the original (1929) version he quoted 200 million million.
[2] See H. Spencer Jones, *Life in Other Worlds*, 1940.

two stars; you know how empty space is, and such an event would be very rare indeed. Of course, what happened once may happen again, but the chance against is very very great. On this problem of life we cannot do better than take the opinion of Sir Arthur Eddington:[1] 'I do not think that the whole purpose of the creation has been staked on the one planet where we live; and in the long run we cannot deem ourselves the only race which has been or will be gifted with the mystery of consciousness. But I feel inclined to claim that *at the present time* our race is supreme; and not one of the profusion of stars in their myriad clusters looks down on scenes comparable to those which are passing beneath the rays of the sun.'

Final Problems. We have now drawn our picture of the Stellar Universe, and formulated a theory of its creation, but the inquiring mind is still unsatisfied. Can we see the whole of the Universe, meaning the region occupied by the spiral nebulae, or shall we some day find more? The 200 in. telescope may perhaps answer this question. Is space infinite? It is difficult to believe that it goes on for ever, but more difficult to answer the question, 'What is at the end of it?' Philosophers suggest that it is boundless but limited in extent, like the surface of a soap bubble. Is it just the Universe, or space itself, that is expanding? Where is it going? They say that the bubble is still being blown. But now we are getting into deep waters and must stop before we are hopelessly lost. Those who would swim on should read the works of Eddington and Jeans.

OBSERVATION

(32) The following objects of interest are shown on the star maps, Figs. 22, 23 and 27, and should be examined with a telescope or binoculars. Objects marked * require a telescope of at least 2 in. aperture, but the others are worth examination with any instrument.

Double stars: Mizar and Alcor, Mizar*, β of the Swan*, Castor*, i of the Crab*, α and β of the Goat.

Nebulae: in Orion and in Andromeda.

Clusters: in Perseus and in the Crab. The Pleiades. The Milky Way.

[1] *Nature of the Physical World*, 1928.

QUESTIONS

Chapter i. Light

1 A. What do you understand by the word 'reflection'?

2 B. State the laws of reflection and describe experiments to illustrate them.

3 A. What kind of mirror is used for (a) an oil or acetylene bicycle lamp, (b) the driving mirror of a car? Give reasons if you can.

4 B. What are the laws of reflection? What would you expect to happen to a number of parallel rays on striking a concave mirror?

5 B. Explain with regard to a concave mirror the meaning of (a) focus, (b) focal length, (c) focal plane, (d) radius of curvature, (e) centre of curvature.

6 B. Describe the images obtainable with a concave mirror, stating clearly the conditions under which each kind occurs.

7 c. An object 1 in. high is placed 6 in. from a concave mirror of focal length 2 in. By making a full-size diagram find the position and size of the image.

8 c. By drawing to scale find the position and size of the image of an object 1 foot high placed 4 feet from a concave mirror of radius 3 feet.

9 B. Explain, with an illustrative experiment, the meaning of 'refraction'.

10 B. Show by diagrams how light passes through (a) a rectangular, (b) a triangular glass block. What rules must the light obey in these cases?

11 B. Show by diagrams (a) refraction, (b) reflection due to a prism. What determines which effect takes place?

12 A. What is a lens? What is it used for?

13 A. How can a lens make things look upside down?

14 B. Explain the meaning of the focus, focal length and focal plane of a convex lens. Give a diagram.

15 B. Explain the difference between real and virtual images.

16C. An object 1 in. high is placed 6 in. from a convex lens of focal length 3 in. Find the position and size of the image.

17C. An object 1 in. high is placed 3 in. from a convex lens of focal length 6 in. Find position and size of the image.

18C. Proceed as in the two previous questions, but place the object 4 in. from a lens of focal length 4 in. and comment on your result.

19A. Why do you sometimes see coloured fringes around the edge of a bevelled mirror?

20B. Distinguish between the deviation and dispersion produced by a prism.

21B. Describe the appearance and production of (a) a continuous spectrum, (b) a line spectrum, (c) an absorption spectrum.

22B. Explain the terms 'ultra-violet' and 'infra-red'.

23B. What is the purpose and action of an achromatic lens?

CHAPTER II. THE EARTH

24A. Have you ever noticed anything that suggests that the Earth is not flat?

25A. What is the effect on the view of climbing a high hill or going up in an aeroplane?

26B. Give an account of the evidence that the Earth is round. Is it a perfect sphere?

CHAPTERS III & IV. ROTATION; CONSTELLATIONS

27A. How does the Sun appear to move during the day? Why?

28A. If the Earth is rotating, why do we not feel it moving?

29A. Are the same stars visible all the year round?

30A. Describe three ways of finding the north.

31B. Give an account of the ancient theory of night and day.

32B. Give a summary of the evidence that the Earth rotates on its axis. How long does one rotation take?

33B. What is meant by the term 'circumpolar stars'? Give a diagram of the more important ones.

34B. How do you recognise the Pole Star?

35B. Name and draw three constellations.

36 B. What constellations occupy the southern part of the sky on an autumn evening? Give diagrams.

37 B. What constellations occupy the southern sky on a winter evening? Give diagrams. Where are the constellations that occupied that part of the sky in autumn?

38 C. What difference would occur in the appearance of the night sky if you travelled to (a) the north pole, (b) China, (c) a place on the equator, (d) the south pole?

CHAPTER V. ANNUAL MOTION

39 A. What other movement has the Earth besides rotation on its axis? How do you know?

40 A. Suggest why Figs. 22 and 23 are different.

41 B. What is meant by 'the Zodiac'? Give a list of the twelve constellations included.

42 B. Draw a diagram, with explanatory notes, to show why the Sun appears to move through certain constellations once a year.

43 B. Explain why Figs. 22 and 23 are different.

44 B. What do you know about the Earth's orbit and the velocity of the Earth in that orbit?

CHAPTER VI. TIME

45 A. Describe the appearance and use of a sundial. Does it agree with an accurate clock?

46 A. What is meant by Greenwich time?

47 A. When summer time comes into force in April do you gain or lose an hour's sleep? What is summer time for?

48 A. How often does February have 29 days? Why?

49 A. What would be meant by '1350 hours' on an Army notice board?

50 A. How could you find the south point during the day and without a compass?

51 B. What is 'the meridian'? Describe a method of finding it.

52 B. Explain the difference between a solar day and a sidereal day.

53 B. Explain the difference between solar and mean time.

54 B. Draw and explain a sundial. Is it accurate? Give reasons for your answer.

55 B. Give an account of the adjustment in time usually spoken of as 'leap year'.

56 B. If a star rises at 8.0 p.m. on a certain day, at what time will it rise a week later? Would this be true of the Sun, reading a.m. for p.m. of course?

57 C. Suggest a method of correctly setting a sundial in your garden, or of testing the setting of an existing one.

58 C. Criticise this statement: 'Lord X was born on February 29, 1894.' Explain any objections that you make.

59 C. Christmas Day 1951 was a Tuesday. What was it in 1950 and 1949, and what will it be in 1953 and 1954? Explain how you found out.

60 C. Study Fig. 36. What has happened to the seconds hand? There certainly was one when the photograph was taken.

CHAPTER VII. POSITION

61 A. The wireless commentaries on the last Test Matches in Australia were given at about breakfast time. Why was this?

62 A. How does a ship in distress give its position when sending out an S.O.S.?

63 A. Why do railway passengers travelling across Europe have to alter their watches at certain places?

64 B. Distinguish between local time and Greenwich Time.

65 B. When it is noon in London, what will be the local time at the following places: Belfast, Long. 6° W.; Calcutta, Long. 88° E.; Wellington, N.Z., Long. 175° E.; Vancouver, Long. 123° W.?

66 B. What is Longitude? Describe a method of finding it.

67 B. What is Latitude? Describe a method of finding it.

68 B. Why is an accurate knowledge of time important to a navigator? Describe the means available for obtaining and keeping it.

69 B. Explain the principle of the sextant.

70B. What corrections have to be applied to a sextant altitude? Give reasons.

71B. Describe the procedure of finding a ship's position at sea.

72C. A man wishing to shoot a bear set off and walked due south for ten miles. Not finding a bear he then turned and walked due east for ten miles. There he shot his bear and then walked exactly ten miles back to his starting point. What was the colour of the bear? Explain your reasoning.

73C. Suggest how the Pole Star could be used as a guide to a long voyage due west.

74C. A man travelled to Switzerland in March, and on the journey made one change in his watch. He returned to London in May and found no such change necessary. Explain this.

75C. In *Round the World in Eighty Days*, by Jules Verne, Phineas Fogg left London on the evening of October 2, travelling eastward, and was due to complete his journey 80 days later, on December 21. He took a few minutes over the 80 days, but found that he had arrived on December 20 and had won his bet. Explain this.

76C. Comment on this portion of a Continental time-table, and explain the apparent delay at Aachen.

Brussels		Dep.	14-40	17-22	17-28
Aachen	G.M.T.	Arr.	17-05	19-42	20-04
	C.E.T.	Dep.	18-37	20-52	21-34
Köln		Arr.	19-32	21-45	22-30

CHAPTER VIII. SEASONS

77A. How does the position in the sky of the summer sun differ from that of winter?

78A. Why is summer sunshine hotter than that of winter?

79A. What causes seasons?

80A. Why is cricket played in Australia at Christmas?

81A. The Earth's axis is said to be tilted. Tilted to what?

82A. What is the 'midnight sun'?

83B. Explain what is meant by the plane of the ecliptic.

84B. How does the tilt of the Earth's axis affect (*a*) temperature, (*b*) the length of the day?

85 B. What are (a) the Equator, (b) the Arctic Circle, (c) the Tropic of Cancer?

86 B. Distinguish between Equinoxes and Solstices. When do they occur?

87 B. How does the direction of sunrise vary throughout the year?

88 C. The south pole is always a cold place, and yet expeditions to those regions usually start about October. Why is this?

89 C. You know the meanings of Arctic Circle and Tropic of Cancer; explain the meanings of Antarctic Circle and Tropic of Capricorn.

90 C. Which are the longest and shortest days (a) in Capetown, (b) at a place on the Equator?

91 C. Answer Question 87, but for a person living in the southern hemisphere.

92 C. Does the changing of the seasons have any effect on the path of the Moon across the sky? (The Moon moves around the Earth in an orbit which lies approximately in the plane of the ecliptic.) Illustrate by diagrams.

CHAPTER IX. THE MOON

93 A. At what time of day and in what direction do you look for (a) a new Moon, (b) a full Moon, (c) a very old Moon?

94 A. Have you ever seen the Moon in the daytime? If so, under what circumstances?

95 A. Which way does the bright portion of a partly illuminated Moon face?

96 A. Does the Moon rotate on its axis or not? Give reasons.

97 A. How would the world be affected if the Moon did not exist?

98 A. What is a shadow? Explain how you would get a sharp shadow, such as you need for making 'bunnies' on the wall with your fingers.

99 A. Can the Moon cast a shadow? Does it ever do so?

100 B. Explain, with a diagram, the cause of the phases of the Moon, and name the various shapes that can be observed.

101 B. Describe how the position of the Moon in the sky varies during the month.

102 B. There are two general types of shadow, due to a point source and an extended source respectively. Explain carefully how they are formed.

103 B. Describe and explain a solar eclipse, mentioning the different types that may occur.

104 B. Describe and explain a lunar eclipse. Why are eclipses so infrequent?

105 B. Write a short essay on 'The Motion of the Moon'.

106 C. What differences, if any, in the appearance of the Moon would you expect to observe if you went to live in New Zealand?

107 C. Since the Sun is very much larger than the Moon, why is it possible for a total eclipse of the Sun to take place?

Chapters X & XI.   Planets; Gravitation

108 A. Certain bodies of star-like appearance seem to move among the stars. What are they called? How many are there? Give the names of some of them.

109 A. Have you ever come across in books, other than astronomical ones, the names which you have given in answer to the last question? If you have, in what connection did they occur?

110 A. Write what you know about 'the evening star'.

111 A. What is the cause of the tides?

112 A. What do you know about (a) Galileo, (b) Sir Isaac Newton?

113 B. Write lists of the nine chief planets, in order of distance from the Sun, the nearest first, and in order of size, the largest first.

114 B. Describe and explain the apparent motion of Venus.

115 B. Describe and explain the apparent motion of Mars.

116 B. Write a short history of the theory of the planets from early times to the seventeenth century.

117 B. Give an account of the discovery of the three planets that were still unknown in the seventeenth century.

118B. Explain the law of gravitation. When and by whom was it discovered?

119B. Explain the cause of the tides. Why do the high tides vary in height during the month?

120B. Draw a diagram of the solar system similar to Fig. 51. The distances can be taken from the table on p. 59; use a scale of 1 mm. = 10 million miles; omit Uranus, Neptune and Pluto.

121C. The next 'transit of Venus' is due in the year 2004. What do you think this phenomenon is? Suggest why we have to wait so long for it.

122C. Write the numbers 0, 3, 6, 12, etc., each double the preceding, until you have eight altogether. To each add 4, making the series 4, 7, 10, etc. Under this series write the distances of the planets from the Sun (p. 59), in tens of millions of miles, i.e. the distance of Mercury being called 3.6. Compare and comment on the two series of numbers. Refer to p. 71 and then reconsider the problem. [This coincidence is called Bode's Law.]

123C. State Kepler's laws of planetary motion. For the Earth and Venus calculate from the table on p. 59 the ratio of the squares of the times, and the ratio of the cubes of the distances, and hence illustrate the third law.

124C. Suggest why the name 'Mercury' has come to be used for the innermost planet, the messenger of the Gods (a classical use), a liquid metal, St Paul (Acts xiv. 12), and in pictorial form as the trademark of National Benzole Mixture.

125C. Study Figs. 43 and 54. At what time would you expect high tide when the Moon was (a) first quarter, (b) full? Suggest factors that might disturb this prediction.

CHAPTER XII. TELESCOPES

126A. Describe a telescope.

127B. Draw a diagram showing the path of two rays through an astronomical telescope.

128B. Explain the action of an astronomical telescope.

129B. Why should a telescope object lens have a long focal length? What is the advantage of a large diameter?

130 B. Explain the action of a Galilean telescope.

131 C. Discuss the advantages and disadvantages of a Galilean telescope as compared with an astronomical.

132 B. Describe the equatorial mounting of an observatory telescope.

133 B. What is an altazimuth mounting? Why is it unsuitable for astronomical photography?

134 B. Write an account of a visit to an astronomical observatory.

135 B. Give an account of the construction and use of a transit instrument.

136 B. One of the routine duties at Greenwich is the determination of correct time. How is it done? Describe carefully the instrument used.

137 B. Describe, with a diagram, the action of a reflecting telescope.

138 B. Describe the methods of mounting reflecting telescopes.

139 B. Explain the meaning of the scales shown on the edges of Fig. 27.

140 C. Use Fig. 27 to describe as accurately as you can the position of a planet in R.A. 10 hours, Dec. 10° N.

141 C. Describe the position of a planet in R.A. 20 h. 30 m., Dec. 20° S. (Fig. 27.)

142 C. What is the R.A. and Dec. of the Pleiades? (Fig. 27.)

CHAPTER XIII. SOLAR SYSTEM

143 B. Write an account of the Moon's surface.

144 B. Describe the inferior planets (i.e. those having orbits smaller than that of the Earth).

145 B. Write what you know of the planet Mars, mentioning any seasonal changes that can be noticed.

146 B. Describe the four great planets.

147 B. Explain why there are variations in the number of visible satellites of Jupiter.

148 B. What changes can be seen in the appearance of Saturn's rings? Why do they occur?

149 B. What are 'the minor planets'? Mention names, and refer to their sizes and orbits.

150 B. Describe the motion and appearance of a comet.

151 B. Explain the cause of 'shooting stars'.

152 B. What are meteorites? Give some account of their origin, motion and composition.

153 B. What is the velocity of light? How can this velocity be determined by astronomical observations?

154 C. Write an account of an imaginary expedition over the surface of the Moon.

155 C. Discuss the possibility of the existence of life elsewhere in the Solar System than on the Earth.

156 C. Obtain a Latin dictionary and prepare a list of English names for the lunar seas shown in Fig. 69.

157 C. Lunar craters are named after famous men. Write what you know about those whose names appear in Fig. 69.

158 C. Examine Fig. 70. Explain why in this drawing the disc is not quite circular.

159 C. Discuss Kepler's laws of planetary motion (p. 55) with regard to the orbit of (a) a planet, (b) a comet.

CHAPTER XIV. THE SUN

160 A. What *is* the Sun? How hot is it?

161 B. What are sunspots? What information about the Sun do they give?

162 B. Describe a total eclipse of the Sun, explaining the various effects that can be seen.

163 B. What are prominences? Under what conditions can they be seen?

164 B. Explain what is meant by the 11-year cycle.

165 B. Describe (a) the aurora, (b) Zodiacal light. What explanations can you suggest?

166 B. Give some account, with possible explanations, of magnetic storms.

167B. Write what you know of the rotation of the Sun.

168B. What are Fraunhofer's lines? Account for their existence in the spectra of the Sun and stars.

Chapter XV. The Stars

169A. Suggest why the stars do not all appear to be of the same brightness.

170A. What relation would you expect to find between brightness and distance?

171A. How far away are the stars?

172B. Explain how the stars are classified according to their brightness.

173B. Describe a method of comparing the brightness of two stars. Write an account of variable stars and novae.

174B. Explain the meaning of the term 'light year'.

175B. Explain what is meant by the 'parallax' of a star.

176B. State and explain the law relating illumination and distance from the source of light.

Chapter XVI. The Stellar Universe

177B. Write a short account of double and binary stars, mentioning the names and positions of at least two examples.

178B. Describe a star cluster. In what constellation may one of these objects be seen? Show by a diagram its exact position.

179B. What is a nebula? Describe two examples of different types.

180B. What is the shape of the star system to which the Sun belongs? Quote evidence in support of your statement.

181B. Explain the term 'extra galactic'.

182B. Criticise that very common expression, 'the fixed stars'.

183B. Give an account of a theory of the evolution of stars and planets.

184B. Discuss the likelihood of the existence of life in other worlds.

BIBLIOGRAPHY

Works on Astronomy are very numerous indeed, and many of them are excellent, though naturally there is a good deal of overlap in their contents. I append here a brief list of books of different types; those marked with an asterisk will probably prove of more value to teachers than to pupils.

GENERAL READING

Sir ROBERT BALL, *The Story of the Heavens* (Cassell). (First published 1886.)

Sir JAMES JEANS, *The Stars in their Courses* (Cambridge).

W. M. SMART, *Astronomy* (Pageant of Progress Series) (Oxford).

OBSERVATIONAL

E. A. BEET. *A Guide to the Sky* (Cambridge).

BERNHARD, BENNETT AND RICE, *New Handbook of the Heavens* (McGraw-Hill).

A. P. NORTON, *New Popular Star Atlas* (Gall and Inglis).

TEXT BOOKS

P. F. BURNS, *First Steps in Astronomy* (Ginn).

*Sir HAROLD SPENCER JONES, *General Astronomy* (Arnold).

E. O. TANCOCK, *Starting Astronomy* (Philip).

MODERN RESEARCH

*Sir ARTHUR EDDINGTON, *The Expanding Universe* (Cambridge).

D. S. EVANS, *Frontiers of Astronomy* (Sigma Books).

*Sir JAMES JEANS, *The Universe Around Us* (Cambridge).

INDEX